AF361016

Seismic Performance Assessment
of Buildings

Seismic Performance Assessment of Buildings

Editors

Rita Bento
Ana Simões

MDPI • Basel • Beijing • Wuhan • Barcelona • Belgrade • Manchester • Tokyo • Cluj • Tianjin

Editors
Rita Bento
University of Lisbon
Portugal

Ana Simões
University of Lisbon
Portugal

Editorial Office
MDPI
St. Alban-Anlage 66
4052 Basel, Switzerland

This is a reprint of articles from the Special Issue published online in the open access journal *Buildings* (ISSN 2075-5309) (available at: https://www.mdpi.com/journal/buildings/special_issues/seismic_performance_assessment_buildings).

For citation purposes, cite each article independently as indicated on the article page online and as indicated below:

LastName, A.A.; LastName, B.B.; LastName, C.C. Article Title. *Journal Name* **Year**, *Volume Number*, Page Range.

ISBN 978-3-0365-2185-5 (Hbk)
ISBN 978-3-0365-2186-2 (PDF)

Contents

About the Editors

Rita Bento (Full Professor), holds a PhD in Civil Engineering and Structures from Instituto Superior Técnico, University of Lisbon and a Master's degree in Earthquake Engineering at Imperial College, London. She is specialized in earthquake engineering, structural behaviour, the dynamic of structures, and the strength of materials. Her main research interests include the assessment, strengthening, and repair of structures; structural testing and nonlinear modelling; earthquake engineering; and structural dynamics. Currently, she is Chairman of the Sub-commission 8 (Eurocode 8) of CT115 (Portuguese Technical Commission for Standardization of Structural Eurocodes). She is also a representative of IPQ [Portuguese entity responsible for quality] at the Commission CEN/TC250/SC8, the director of the PhD program of IST Analysis and Mitigation of Risks in Infrastructures, a member of Executive Board of The European Association of Earthquake Engineering (EAEE), an associated Editor of Bulletin of *Earthquake Engineering*, coordinator of the EAEE Work Group 12 (WG12: Continuing Education and Professional Development), and represents in Portugal software 3Muri from S.T.A. DATA.

Ana Simões (PhD), holds a MSc and PhD degree in Civil Engineering from Instituto Superior Técnico, University of Lisbon. She is the co-author of several publications in scientific journals, conference proceedings, and books. She develops research activity at Civil Engineering Research and Innovation for Sustainability (CERIS) in the field of the seismic performance assessment of unreinforced masonry buildings. She also represents in Portugal software 3Muri from S.T.A. DATA and carries out design activity at Barraferros, Lda.

buildings

Editorial

Seismic Performance Assessment of Buildings

Rita Bento * and Ana Simões

CERIS, Instituto Superior Técnico, University of Lisbon, 1049-001 Lisbon, Portugal;
ana.g.simoes@tecnico.ulisboa.pt
* Correspondence: rita.bento@tecnico.ulisboa.pt

Citation: Bento, R.; Simões, A.
Seismic Performance Assessment of
Buildings. *Buildings* **2021**, *11*, 440.
https://doi.org/10.3390/
buildings11100440

Received: 22 September 2021
Accepted: 26 September 2021
Published: 28 September 2021

The seismic performance assessment of buildings is a challenging process. Despite the valuable guidelines in force for the seismic assessment of buildings, including the Eurocode 8 part 3 [1] and the Italian Technical Code [2,3] in Europe, FEMA 273/274 [4] in the US, and the New Zealand Society of Earthquake Engineering (NZSEE) Seismic Assessment guidelines [5] in New Zealand, several open issues require attention from the scientific community. This Special Issue provides an overview of the present knowledge related to various aspects of "Seismic Performance Assessment of Buildings" and identifies further studies needed concerning this area of research. A total of seven original research studies were published, with relevant contributions from international experts from Italy, Portugal, Switzerland, and New Zealand. The contributions address the seismic assessment of rein-forced concrete (RC) wall buildings, RC buildings at urban scale, residential unreinforced masonry (URM) buildings, including timber diaphragms and structural irregularities, complex monumental URM buildings, and adobe buildings.

Orumiyehei and Sullivan [6] propose a newly simplified probabilistic displacement-based assessment approach for RC wall buildings. The approach is a simplification of the traditional displacement-based assessment approach as it reduces the steps required to compute damping and spectrum reduction factor, and it is converted into a probabilistic approach by accounting for pertinent variabilities to assess the building's seismic risk in terms of the annual probability of exceeding key limit states. To evaluate the accuracy and limitations of the proposed approach, a series of RC buildings with 4-, 8-, and 12-stories were assessed by comparing the assessed likelihood of exceeding key limit states obtained from the simplified and rigorous (multi-stripe analyses) probabilistic approach. The results indicate that the former approach provides good estimates of the median intensity associ-ated with exceeding a given failure mechanism and the annual probability of exceeding limit states. This newly simplified probabilistic displacement-based assessment approach for RC wall buildings is perceived as a valuable extension to the displacement-based assessment method considered in the current assessment guidelines in New Zealand [5].

Flora et al. [7] proposes a novel simplified approach for the seismic loss assessment of RC buildings at urban scale based on the direct estimation of expected annual loss (DEAL) method, which aims to provide a general understanding of the socio-economic impacts of seismic scenarios at a territorial scale. The simplified approach was applied to the residential building stock of two main areas of the center of Potenza (Italy). Non-linear static analyses (NLSA) were carried out in a series of case-study buildings (archetypes), each one associated with the main RC structural typologies identified, to derive the main engineering demand parameters (EDPs), which represent the input data for the application of the DEAL method and the estimation of the EAL (expected annual loss) of RC buildings. The EAL associated with indirect losses related to downtime was also estimated based on the method presented in Cardone et al. [8]. With the results in terms of monetary losses, preliminary considerations on the economic and social impacts of probable seismic scenarios were made.

Aşıkoğlu et al. [9] present an overview of the NLSA procedures within the framework of the performance-based approach for URM with structural irregularity. The authors highlight two major issues: (1) the lack of a systematic and uniform procedure for defining structural irregularities in masonry buildings and (2) the applicability of classical NLSA procedures to irregular masonry buildings, as these were developed for regular buildings. Two masonry case studies with different irregularity levels were selected from the literature to exemplify the application of different NLSA. It was demonstrated that the maximum displacement demand was highly dependent on the empirical formulations, stressing the concerns about the application of both classical and extended (improved so that the effects of irregularities on the response can be included) NLSA procedures. The main purpose of the review was to point out the complexity of structural irregularities and their influence on the procedures adopted to achieve performance limits and to show, to a great extent, that further studies are still needed concerning the applicability of NLSA procedures to irregular masonry buildings.

Guerrini et al. [10] compare different methods for calculating earthquake-induced displacement demands associated with NLSA procedures to assess URM buildings. The methods considered in this study are divided into two main families: methods based on the concept of equivalent linear systems and methods that employ inelastic response spectra. First, the authors discuss the accuracy of two established methods per family, highlighting their main shortcomings, and then present an improved formulation for each family, the optimal stiffness method (OSM) and the modified N2 method (MN2 method). The accuracy of the improved formulations was assessed based on the results from non-linear time-history analyses (NLTHA), carried out on single-degree-of-freedom oscillators with hysteretic force–displacement relationships representative of URM buildings. It was concluded that both proposed formulations predict the median ductility demand accurately while limiting the dispersion of the results.

Tomić et al. [11] explore the seismic assessment of URM buildings with timber diaphragms by explicitly modeling the diaphragm stiffness and the finite strength of wall-to-diaphragm connections through a newly developed equivalent frame macro-element proposed by Vanin et al. [12] able to simulate both in-plane and out-of-plane behavior of URM buildings. The modeling approach was used to model an unstrengthened stone masonry building, experimentally investigated through a shake table (Pavia Building 1 [13,14]), and to simulate three retrofit interventions: (1) diaphragms retrofitted with an additional layer of timber planks, (2) retrofitted wall-to-diaphragm connections, and (3) the combination of both interventions. Based on these simulations (unstrengthened and strengthened), the authors validated the new modeling approach for the seismic assessment of URM buildings with unstrengthened timber floors and showed that strengthening the timber diaphragm alone is ineffective when the friction capacity of the wall-to-diaphragm connection is exceeded. This puts in evidence the importance of explicitly modeling the diaphragm stiffness and the finite strength of wall-to-diaphragm connections if the equivalent frame model captures both global in-plane and local out-of-plane failure modes.

Lagomarsino et al. [15] address the seismic assessment of the Podestà Palace in Mantua (Italy), highlighting the main issues with the assessment of complex monumental URM buildings composed of various units stratified over centuries. The authors propose an integrated use of three modeling strategies characterized by a different computational effort and degree of accuracy to: (1) assess the global response of the whole structure and estimate the mutual dynamic interactions among the structural units, (2) assess the out-of-plane response of facades prone to the activation of local mechanisms, and (3) deepen the seismic response of some critical parts. This integrated approach uses the results achieved from one modeling approach as input for another (e.g., the floor spectra estimated by (1) were used to define the seismic input in (2)). This aimed to get a comprehensive interpretation of the seismic behavior of Podestà Palace and to address more rationally possible strengthening solutions, parametrically investigated and specifically conceived for the safety and preservation of the monument.

Momin et al. [16] approach the seismic vulnerability assessment of Portuguese adobe buildings as part of the endeavor towards the preservation of the inheritance and cultural heritage of the country. Three buildings with one-story, two-stories, and two-stories plus an attic were numerically modeled using solid and contact elements. NLTHA was performed until complete collapse occurred. Two novel EDPs were used: the crack propagation ratio (CPR), which refers to the cracks that develop in the walls prior to the onset of the detachment of the blocks, and the building volume loss ratio (VLR), which refers to the post-failure movements of the blocks that have the potential to cause fatalities. According to the authors, the choice of the building VLR is considered a better damage descriptor for estimating risk to occupants as compared to traditional damage states because it can be directly correlated with earthquake fatalities. The authors also proposed damage thresholds in correlation with the damage classifications of the European Macroseismic Scale (EMS-98) [17]. The seismic vulnerability assessment was concluded with the derivation of fragility functions using cloud analysis to quantify physical, structural damage due to a given intensity measure (IM). In this case, the peak ground acceleration (PGA) was chosen for fatality vulnerability functions to estimate indoor fatalities.

The editors would like to acknowledge all authors for their valuable contribution in different fields of knowledge related to "Seismic Performance Assessment of Buildings". The editors express their gratitude to the peer reviewers for their rigorous analysis of the different contributions and the managing editors of *Buildings* involved in this Special Issue.

Author Contributions: Conceptualization, R.B. and A.S.; writing—original draft preparation, R.B. and A.S.; writing—review and editing, R.B. and A.S. Both authors have read and agreed to the published version of the manuscript.

Funding: This research received no external funding.

Institutional Review Board Statement: Not applicable.

Informed Consent Statement: Not applicable.

Conflicts of Interest: The authors declare no conflict of interest.

References

1. CEN. *Eurocode 8: Design of Structures for Earthquake Resistance-Part 1: General Rules, Seismic Actions and Rules for Buildings*; CEN: Brussels, Belgium, 2005.
2. NTC 2018. Italian Technical Code, Decreto Ministeriale 17/1/2018. Aggiornamento Delle Norme Tecniche per le Costruzioni. Ministry of Infrastructures and Transportation, G.U. n.42 of 20/2/2018. Available online: https://www.studiopetrillo.com/ntc2018.html (accessed on 20 September 2021). (In Italian)
3. Ministry of Infrastructures and Transportation. C.S.Ll.PP. n.7 del 21/01/2019. Istruzioni per L'Applicazione Dell'Aggiornamento delle Norme Tecniche per le Costruzioni di cui al D.M. 17/01/2018 G.U. S.O. n.35 of 11/2/2019. Available online: https://www.gazzettaufficiale.it/eli/id/2019/02/11/19A00855/sg (accessed on 20 September 2021). (In Italian).
4. Federal Emergency Management Agency (FEMA). *NEHRP Guidelines for the Seismic Rehabilitation of Buildings and Commentary*; FEMA: Washington, DC, USA, 1997; pp. 273–274.
5. New Zealand Society of Earthquake Engineering (NZSEE). *The Seismic Assessment of Existing Buildings: Technical Guidelines for Engineering Assessments*; Part C—Detailed Seismic Assessment; NZSEE: Wellington, New Zealand, 2017.
6. Orumiyehei, A.; Sullivan, T.J. Displacement-Based Seismic Assessment of the Likelihood of Failure of Reinforced Concrete Wall Buildings. *Buildings* **2021**, *11*, 295. [CrossRef]
7. Flora, A.; Cardone, D.; Vona, M.; Perrone, G. A Simplified Approach for the Seismic Loss Assessment of RC Buildings at Urban Scale: The Case Study of Potenza (Italy). *Buildings* **2021**, *11*, 142. [CrossRef]
8. Cardone, D.; Flora, A.; Picione, M.D.L.; Martoccia, A. Estimating direct and indirect losses due to earthquake damage in residential RC buildings. *Soil Dyn. Earthq. Eng.* **2019**, *126*, 105801. [CrossRef]
9. Aşıkoğlu, A.; Vasconcelos, G.; Lourenço, P.B. Overview on the Nonlinear Static Procedures and Performance-Based Approach on Modern Unreinforced Masonry Buildings with Structural Irregularity. *Buildings* **2021**, *11*, 147. [CrossRef]
10. Guerrini, G.; Kallioras, S.; Bracchi, S.; Graziotti, F.; Penna, A. Displacement Demand for Nonlinear Static Analyses of Masonry Structures: Critical Review and Improved Formulations. *Buildings* **2021**, *11*, 118. [CrossRef]
11. Tomić, I.; Vanin, F.; Božulić, I.; Beyer, K. Numerical Simulation of Unreinforced Masonry Buildings with Timber Diaphragms. *Buildings* **2021**, *11*, 205. [CrossRef]
12. Vanin, F.; Penna, A.; Beyer, K. A three-dimensional macroelement for modelling the in-plane and out-of-plane response of masonry walls. *Earthq. Eng. Struct. Dyn.* **2020**, *49*, 1365–1387. [CrossRef]

13. Magenes, G.; Penna, A.; Galasco, A. A full-scale shaking table test on a two-storey stone masonry building. In Proceedings of the 14th European Conference on Earthquake Engineering, Ohrid, North Macedonia, 30 August–3 September 2010.
14. Magenes, G.; Penna, A.; Senaldi, I.E.; Rota, M.; Galasco, A. Shaking table test of a strengthened full-scale stone masonry building with flexible diaphragms. *Int. J. Archit. Herit.* **2014**, *8*, 349–375. [CrossRef]
15. Lagomarsino, S.; Degli Abbati, S.; Ottonelli, D.; Cattari, S. Integration of Modelling Approaches for the Seismic Assessment of Complex URM Buildings: The Podestà Palace in Mantua, Italy. *Buildings* **2021**, *11*, 269. [CrossRef]
16. Momin, S.; Lovon, H.; Silva, V.; Ferreira, T.M.; Vicente, R. Seismic Vulnerability Assessment of Portuguese Adobe Buildings. *Buildings* **2021**, *11*, 200. [CrossRef]
17. Comisión Sismológica Europea. Escala Macro Sísmica Europea EMS-98. 1998, Volume 15. Available online: http://media.gfz-potsdam.de/gfz/sec26/resources/documents/PDF/EMS-98_Original_englisch.pdf (accessed on 20 September 2021).

buildings

Article

Displacement-Based Seismic Assessment of the Likelihood of Failure of Reinforced Concrete Wall Buildings

Amirhossein Orumiyehei and Timothy J. Sullivan *

Civil and Natural Resources Engineering, University of Canterbury, Christchurch 8041, New Zealand; amir.orumiyehei@pg.canterbury.ac.nz
* Correspondence: timothy.sullivan@canterbury.ac.nz

Abstract: To strengthen the resilience of our built environment, a good understanding of seismic risk is required. Probabilistic performance-based assessment is able to rigorously compute seismic risk and the advent of numerical computer-based analyses has helped with this. However, it is still a challenging process and as such, this study presents a simplified probabilistic displacement-based assessment approach for reinforced concrete wall buildings. The proposed approach is trialed by applying the methodology to 4-, 8-, and 12-story case study buildings, and results are compared with those obtained via multi-stripe analyses, with allowance for uncertainty in demand and capacity, including some allowance for modeling uncertainty. The results indicate that the proposed approach enables practitioners to practically estimate the median intensity associated with exceeding a given mechanism and the annual probability of exceeding assessment limit states. Further research to extend the simplified approach to other structural systems is recommended. Moreover, the research highlights the need for more information on the uncertainty in our strength and deformation estimates, to improve the accuracy of risk assessment procedures.

Keywords: seismic risk; probability of failure; seismic assessment; probabilistic displacement based assessment; reinforced concrete wall

Citation: Orumiyehei, A.; Sullivan, T.J. Displacement-Based Seismic Assessment of the Likelihood of Failure of Reinforced Concrete Wall Buildings. *Buildings* **2021**, *11*, 295. https://doi.org/10.3390/buildings11070295

Academic Editors: Rita Bento, Ana Simões and A. Athanatopoulou-Kyriakou

Received: 23 April 2021
Accepted: 1 July 2021
Published: 6 July 2021

Publisher's Note: MDPI stays neutral with regard to jurisdictional claims in published maps and institutional affiliations.

1. Introduction

Guidelines for the effective quantification and mitigation of seismic risk are essential for the resilience of our built environment and communities in general. A number of valuable guidelines already exist for the seismic assessment of buildings, including the Eurocode 8 part 3 [1] in Europe, FEMA 273/274 [2] in the US, and the MBIE NZSEE Seismic Assessment guidelines [3] in New Zealand. Such guidelines, however, tend to focus on quantifying whether an existing building can or cannot withstand a particular intensity of shaking, without quantifying the seismic risk rigorously, considering the uncertainty in demand and capacity to identify the likelihood of different damage levels (e.g., collapse) occurring. Building owners and communities that are provided with better information on seismic risk are likely able to implement more effective measures to reduce the seismic risk and improve resilience.

A number of proposals have been made in the literature to permit seismic risk assessment in a simplified manner, including the works of [4–10]. Such simplified methods are desirable since they are more likely to be widely implemented in practice. However, those methods that rely on linear elastic analysis are unable to accurately account for the effect of non-linearity in the system. Methods that utilize non-linear static (pushover) analyses resolve this issue, but do rely on the engineer having access to suitable software and having the skills required to develop an accurate non-linear model and execute the analysis properly. Furthermore, traditional non-linear static analysis does not account for the effects of higher modes and, hence, additional measures are required for the engineer to capture these effects. As an alternative to software-driven pushover analyses, Priestley et al. [11] advocates a displacement-based assessment procedure that has been implemented within the New Zealand national seismic assessment guidelines [3]. The approach can be used to establish, in a simplified fashion, the force-displacement (pushover) curve for a structure, and there are a number of simplified expressions that can be used to assess the impact of higher mode effects on the system. Nevertheless, as pointed out by Orumiyehei and Sullivan [12], the traditional DBA procedure of Priestley et al. [11] does not include the effect of uncertainties in quantifying the annual probability of exceeding assessment limit states. To address this, Orumiyehei and Sullivan [12] proposed a modified displacement-based assessment procedure. In this paper, the probabilistic DBA approach of Orumiyehei and Sullivan [12] will be extended to the case of RC wall buildings. By virtue of this study, the traditional displacement-based assessment approach for multi-story wall buildings is simplified by reducing the steps required to compute damping and spectrum reduction factor, and converted into a probabilistic approach by accounting for the pertinent variabilities. As part of the research, new relationships between intensity and ductility demands will also be proposed. The performance of the proposed methodology will be gauged by comparing the assessed likelihood of exceeding key limit states with values obtained from rigorous probabilistic assessment methods.

1.1. Probabilistic Displacement Based Assessment

An overview of the procedure that is proposed to undertake probabilistic displacement-based seismic assessment is provided in Figure 1, from Orumiyehei and Sullivan [12]. This procedure has been developed to provide engineers with a simplified means of quantifying the annual probability of exceeding a limit state of interest. The first step in the seismic assessment of a building will typically require a review of drawings (if available) and a visit to the site to identify the lateral load resisting system (and its condition), including the floor diaphragms, likely material properties, and the characteristics of potentially vulnerable non-structural elements. The first calculation step will then seek to identify the likely plastic mechanism, as shown in Figure 1a. This is used, together with structural and non-structural deformation and strength limits, to define the expected displaced shape of the building at attainment of the limit state, illustrated in Figure 1b. With the displaced shape and lateral resistance of the expected mechanism established, the characteristics of an equivalent single-degree-of-freedom (SDOF) system, such as that shown in Figure 1c, can be computed and a force-displacement capacity curve defined, as illustrated in Figure 1d. These four steps of the procedure are in line with the procedure of Priestley et al. [11] and have been verified quite extensively in the realm of both design and assessment (see, for example, Priestley et al. [11], Sullivan and Calvi [13], Cardone et al. [14], Pampanin et al. [15]). However, the next step uses the limit state displacement capacity, Δ_{cap}, together with new empirical relationships between intensity and displacement (ductility) demands, to compute the median spectral acceleration capacity, $S_{a,cap}$, as per Figure 1e.

The initial period of vibration of the structure, T_i, is then estimated. This can be done using the equivalent SDOF mass, m_e, yield displacement, Δ_y, and yield strength, V_y, as per Equation (1):

$$T_i = 2\pi \sqrt{\frac{m_e \cdot \Delta_y}{V_y}}, \tag{1}$$

The building's initial period and spectral acceleration capacity can then be used, together with information on the intensity of different return periods (i.e., the seismic hazard), to establish the median return period intensity, T_R, for a limit state, as per Figure 1f. Finally, the annual probability of exceeding the limit state, P_{LS}, can be computed via Equation (2), proposed by Vamvatsikos [16].

$$P_{LS} = \sqrt{p}\, k_0^{1-p} [1/T_R]^p \exp\left[\frac{k_1^2}{4k_2}(1-p)\right], \tag{2}$$

where k_0, k_1, and k_2 are coefficients used to describe the hazard as the mean annual frequency (MAF) at which different levels of shaking intensity, S_a, are exceeded at a site, assuming the second-order expression proposed by Vamvatsikos [16] and shown as Equation (3), and p is given by Equation (4).

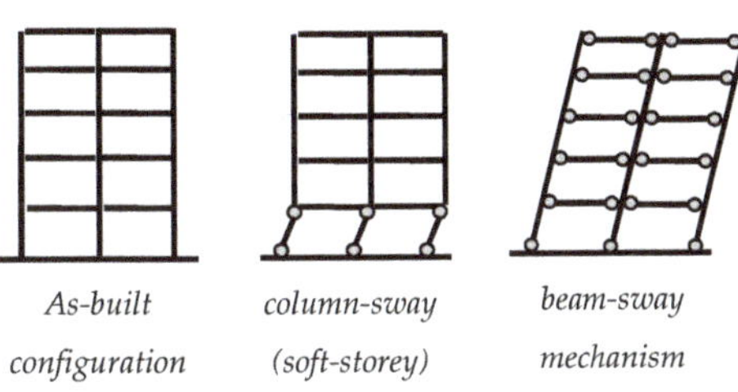

(a) Identification of the expected plastic mechanism

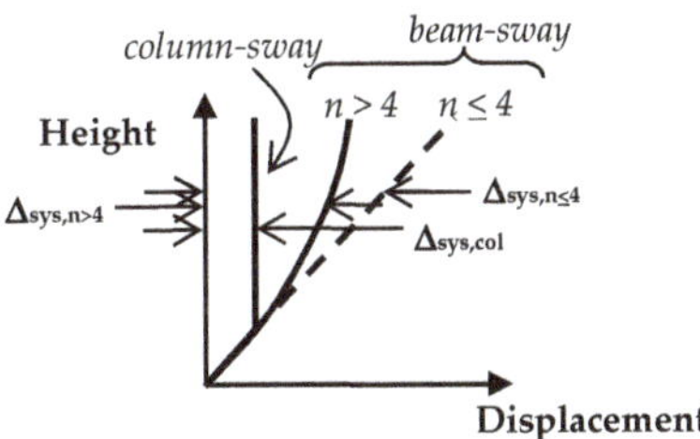

(b) Estimation of the displaced shape at the limit state

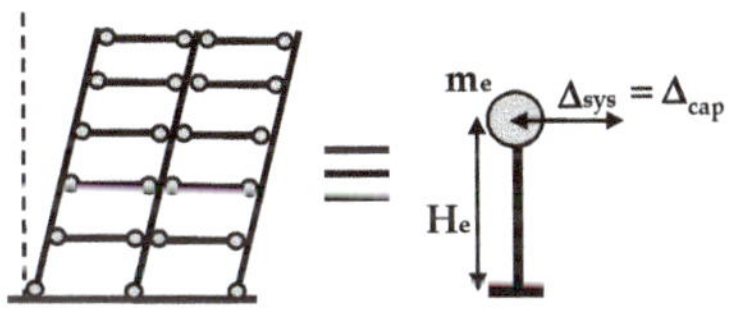

c) Transformation into equivalent SDOF system

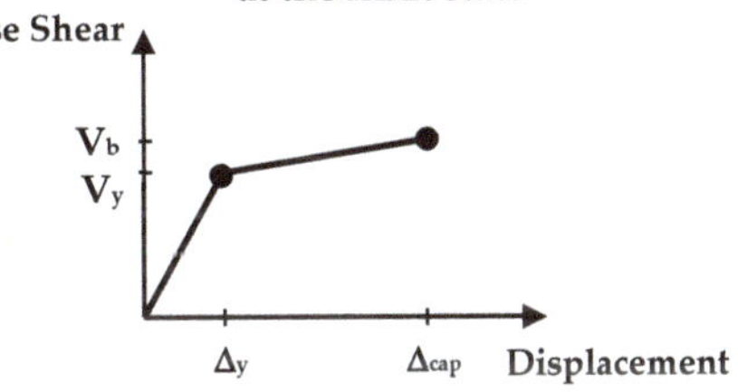

(d) Definition of pushover (force-displacement) response curve

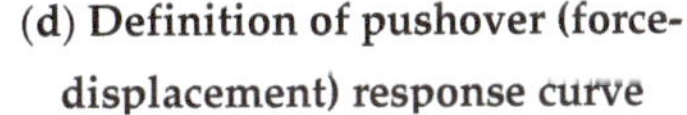

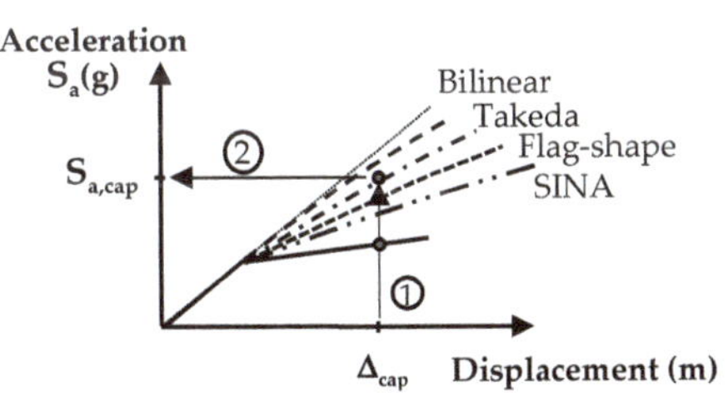

(e) Identification of spectral acceleration capacity associated with displacement

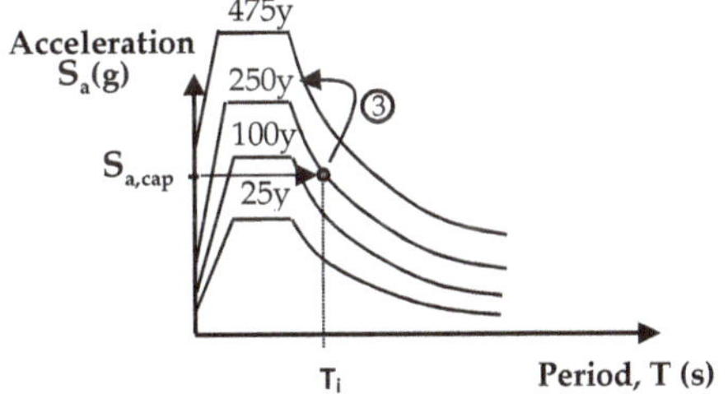

(f) Identification of median return period intensity that generates limit state

Figure 1. Overview of probabilistic displacement-based assessment approach, Adapted with permission from [12]. Copyright 2021 Journal of Earthquake Engineering.

$$\mathrm{MAF} = k_0 \exp(-k_2 \ln^2(S_a) - k_1 \ln S_a), \tag{3}$$

$$p = \frac{1}{1 + 2k_2\left(\beta_{Tot}^2\right)}, \tag{4}$$

where β_{Tot} is the total uncertainty, which Bakalis and Vamvatsikos [17] recommend be computed considering the dispersion in demand, β_D, and capacity, β_c, as per Equation (5).

$$\beta_{Tot} = \sqrt{\beta_D^2 + \left(\frac{\beta_c}{b}\right)^2}, \tag{5}$$

where b is a coefficient that is a function of the hysteretic characteristics of the building and is used in the empirical relationship between seismic intensity and displacement (ductility) demand, as will be explained in detail in Section 1.2.

The information required for the last part of this assessment procedure (Equations (2)–(5) above) is likely to create some difficulty for practitioners. To overcome this, ideally, the coefficients of the hazard curve (k_0, k_1, and k_2) could be provided as part of national seismic hazard information. Furthermore, assessment guidelines could provide information on typical values of dispersion in demand and capacity, or directly provide the value of β_{Tot} (which will typically be between 0.4 and 0.6).

In this paper, the approach described above is extended and tested for multi-story RC wall systems. Furthermore, improved relationships between seismic intensity (referred to as an intensity measure, IM) and displacement ductility demand (an engineering demand parameter, EDP) is developed as explained in the next sub-section.

1.2. Empirical Relationships between Seismic Intensity and Displacement Ductility Demand

The proposed displacement-based assessment method relies on the use of relationships between displacement demand and seismic intensity (as shown in Figure 1e and through the b-value indicated in Equation (7)). To this extent, there exists a variety of ways to relate seismic intensity with displacement (ductility) demand. For example, the equal-displacement rule [18] suggests that ductility demands increase in proportion to the seismic intensity. Generally speaking, methods to relate intensity with engineering demand parameters (such as displacement ductility) could be divided into two main groups. Firstly, comprehensive methods that employ numerical solutions, such as incremental dynamic analysis (IDA) [19] and multi stipe analysis (MSA) [20]. Secondly, approaches that aim to simplify the process of engineering demand estimation; among the second group one could include the capacity spectrum approach [21], the SAC/FEMA approach [9], the N2 method [22], and the displacement based design/assessment approach [11].

In the well-known SAC/FEMA framework [9,23,24], Equation (6) is recommended as a simplistic and practical IM-EDP relationship.

$$\Delta = a(S_a)^b,\qquad(6)$$

where Δ is the displacement demand, S_a is the spectral acceleration intensity and a and b are scaling coefficients.

The FEMA350 [23] guidelines recommend that the coefficient b is set according to the equal-displacement rule [18], with b = 1.0. However, there have been many studies showing that the equal-displacement rule is inaccurate for certain period ranges and hysteretic models [8,25]. To this extent, Jalayer [26] do propose that the coefficients in Equation (6) be set by conducting regression analysis of nonlinear time history analysis results. Orumiyehei and Sullivan [12] conducted such regression analysis and arrived at a new set of b values for a range of periods and hysteresis models. However, it is noted here that for any value of b that is not unity, the units of Equation (6) are unrealistic.

In light of the above, the following model is proposed to relate seismic intensity, S_a, and displacement ductility demand, μ.

$$\frac{S_a}{S_{ay}} = 1 + (\mu - 1)^{1/b},\qquad(7)$$

where S_{ay} is the spectral acceleration at yield and b is an empirical regression coefficient that is unitless (overcoming the issue with the coefficient b indicated in Equation (6)). The coefficient b can then be seen as an indicator of the rate at which inelastic displacement demands increase as spectral acceleration demands increase, as illustrated in Figure 1e. Note that the seismic intensity is expressed in spectral acceleration. While the spectral velocity or spectral displacement demand could be used via conversion to spectral acceleration demands, the use of other intensity measures, such as peak ground acceleration or arias intensity, would require new relationships between intensity and displacement (ductility) demands to be developed.

The regression process described in [12] has been repeated herein with the new model function given by Equation (7), using results of non-linear time-history analyses described in Stafford et al. [27]. This lead to the b-values, together with the associated dispersion, reported in Table 1. Observe that for the bilinear model at medium periods the b values are close to 1.0, which would correspond to the equal-displacement rule. However, for other hysteretic models and periods it can be seen that much larger values of b are obtained. Note that a generalized value of b at short and medium range periods are presented in Table 2.

Table 1. Median b values obtained from regression, as a function of period and hysteretic model, for use in Equation (7).

Period	Bilinear		Takeda		Flag (λ = 5.67)		SINA	
T(s)	$\hat{b}$	β	$\hat{b}$	β	$\hat{b}$	β	$\hat{b}$	β
0.	3.10	0.08	5.55	0.27	5.65	0.21	5.89	0.22
0.2	1.54	0.02	2.16	0.08	2.87	0.04	4.05	0.21
0.3	1.30	0.04	1.72	0.12	2.05	0.20	2.45	0.24
0.4	1.24	0.05	1.49	0.08	1.86	0.18	2.63	0.26
0.5	1.18	0.05	1.44	0.09	1.73	0.15	2.04	0.24
0.6	1.14	0.07	1.36	0.09	1.62	0.11	1.79	0.16
0.8	1.09	0.05	1.29	0.07	1.58	0.07	1.71	0.16
1.0	1.10	0.08	1.23	0.06	1.51	0.07	1.57	0.12
1.5	1.12	0.10	1.20	0.12	1.37	0.06	1.31	0.10
2.0	1.10	0.07	1.23	0.07	1.47	0.10	1.30	0.07
2.5	1.17	0.10	1.24	0.07	1.45	0.09	1.26	0.09
3.0	1.24	0.14	1.28	0.11	1.48	0.14	1.33	0.12

$\hat{b}$ indicates the median of the b-values assuming a lognormal distribution; β represents the associated dispersion, λ is used to characterize the flag-shape hysteretic model; refer to [12].

Table 2. Generalization of the proposed 'b' value to short and medium period ranges.

Period	Bilinear		Takeda		Flag (λ = 5.67)		SINA	
T(s)	b	β	b	β	b	β	b	β
$0.2 \leq T < 0.6$	1.28	0.12	1.57	0.24	1.86	0.34	2.50	0.42
$0.6 \leq T \leq 3.5$	1.12	0.10	1.26	0.12	1.49	0.12	1.38	0.17

2. Assessing the Likely Failure Mechanism for a RC Wall Building

As part of the displacement-based seismic assessment process described in Section 1, the likely failure mechanism should be identified. This could be done through pushover analysis, provided that all types of mechanisms and effects can be adequately modeled. Alternatively, Priestley and Calvi [28] and Priestley et al. [11] provide a hand-calculation procedure in which the relative strengths of different elements and actions are compared to identify the most likely yielding sequence. For a RC wall building, this process is easiest done by first assuming that a flexural plastic hinge will be able to form at the base of the RC walls. Subsequently, the internal shear forces associated with this mechanism can then be computed and checks made to establish whether other mechanisms are more likely. The following section will illustrate how the likelihood of two other possible mechanisms can be checked, wall shear-mechanisms, and foundation overturning mechanisms. Whilst out of scope for this paper, one should also check that floor diaphragms are sufficiently strong to transfer inertia forces between the floors and the walls.

2.1. Flexural Plastic Hinging or a Wall Shear-Mechanism

To establish whether a wall is more likely to develop a flexural mechanism with plastic hinging at the wall base or a shear failure, the expected flexural strength of the base wall section, M_n, is first calculated. To do this, one should use traditional RC section analysis approaches, but with expected (as opposed to characteristic) yield strength values for the reinforcing steel and concrete. Priestley et al. [11] suggest that in-lieu of more accurate data on the materials, one could take the expected strength of reinforcing steel and concrete as being 1.1 and 1.3 times the characteristic values, respectively.

As shown in Equation (8), the wall base flexural strength can be divided by the effective height (H_e) of an equivalent SDOF representation of the building to obtain the equivalent base shear force (V_b) required to cause flexural yielding.

$$V_b = M_n / H_e, \tag{8}$$

where the effective height is computed as:

$$H_e = \frac{\sum m_i \Delta_i h_i}{\sum m_i \Delta_i}, \tag{9}$$

where m_i is the mass of level i, h_i is the height above the base hinge to level i and Δ_i is the lateral displacement of level i of the wall system at the limit state of interest. The summations are done considering all levels in the building.

With the objective of identifying which mechanism is likely to form, the displacement profile to be used in Equation (9) can be taken as the lateral displacement profile at the point of wall-yield, $\Delta_{y,i}$, as [11]:

$$\Delta_{y,i} = \frac{\phi_y h_i^2}{2} \left(1 - \frac{h_i}{3H} \right), \tag{10}$$

where h_i is the height of level i above ground, H is the total height of the wall, and ϕ_y is the nominal yield curvature (ϕ_y) of the wall base, which can be found either from moment-curvature section analyses or via Equation (11) (from [11]):

$$\phi_y = \frac{e \varepsilon_y}{D}, \tag{11}$$

where D is the section depth, corresponding to the wall length, L_w, in the case of RC walls, ε_y is the yield strain of the longitudinal reinforcement and e is an empirical constant that varies according to the section shape. The value for e is 2.0 for rectangular sections [11] and 1.4 for U-shaped and I-shaped sections bending parallel to the web (see [29]).

With the base shear obtained, a set of equivalent lateral forces, F_i, can then be found via Equation (12):

$$F_i = \frac{m_i \Delta_i}{\sum m_i \Delta_i} V_b, \tag{12}$$

where all the symbols have been defined above and where again, for the purposes of identifying the lateral mechanism, Δ_i can be taken as the yield displacement profile for the wall system (via Equation (10)).

By summing the equivalent lateral forces down from roof level, one obtains the story shear demands, which in turn can be used to obtain the shear and bending moment demands on individual walls. Thus, at this stage one is able to identify the effective first mode displacement, shear force and bending moment profiles for the walls, as illustrated in Figure 2.

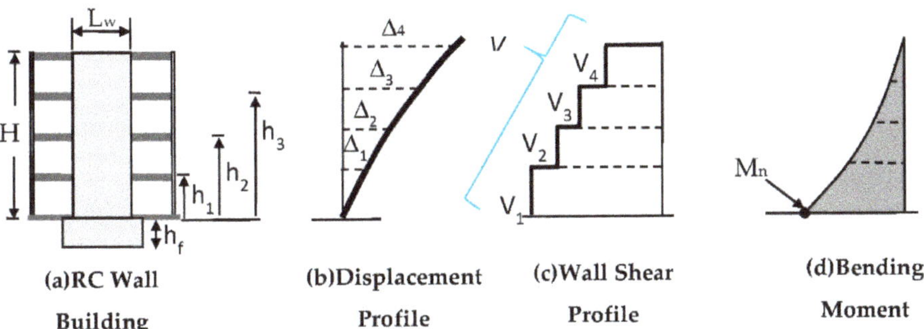

(a)RC Wall Building   **(b)Displacement Profile**   **(c)Wall Shear Profile**   **(d)Bending Moment**

Figure 2. Equivalent first mode lateral displacement, shear, and moment diagrams for a cantilever RC wall building.

The peak shear forces expected in the walls will need to consider all relevant modes of vibration in the building. To this extent, the equivalent first mode shears identified following the procedure described above can be amplified to account for higher-mode effects. There are various proposals in the literature to account for higher mode effects on wall shear demands. For example, Pennucci et al. [30] and Fox et al. [31] show that the amplification will depend on the intensity of shaking and the level of ductility demand in the walls. Standards, such as Eurocode 8 [1] and the NZ standard [32], include more simplified expressions for maximum expected wall shear demands ($V_{i,max}$). For example, NZS 3101 [32] provides the following expression for maximum expected shear demands

$$V_{i,max} = \varphi_o \omega_v V_{i,E}, \tag{13}$$

where $V_{i,E}$ is the wall shear expected for level i considering the base flexural strength (i.e., the shear profile shown in Figure 2) multiplied by the plastic hinge overstrength, φ_o, and ω_v is a dynamic magnification factor (to account for higher modes) given by:

$$\omega_v = 1.3 + \frac{n}{30} < 1.8, \tag{14}$$

where n is the number of floors.

After amplifying the first mode shear force profile to account for higher mode effects, the maximum shear force demands in the walls can then be compared to the wall shear force capacity. If the shear capacity is less than the shear force demands, it is concluded that a shear mechanism is more likely than a flexural mechanism, and vice versa.

2.2. Flexural Plastic Hinging or Foundation Overturning?

Another possible mechanism that should be checked is foundation overturning or rocking. The overturning demand, M_{OT}, associated with flexural hinging at the base of the wall can be found as:

$$M_{OT} = M_{wall,Ov} + V_b h_f, \tag{15}$$

where V_b is the peak base shear force, $M_{wall,Ov}$ is the overstrength flexural resistance of the base plastic hinge and h_f is the foundation height (see Figure 2). Note that when computing the peak overturning demand allowance for higher mode effects on shear demands can be made as per the previous section, but this should have limited effect on the overturning demands.

The overturning demand obtained from Equation (15) can be compared with the overturning capacity of the foundation, assessed considering the expected axial load and shear demand on the wall, and the ground conditions. Sullivan et al. [33] and Millen et al. [34] provide guidance on displacement-based assessment considering soil-foundation-structure-interaction. However, in principle any accepted approach for the computation of foundation overturning capacity could be adopted. If the assessed overturning capacity of the foundation is less than the demand given by Equation (15), then a foundation mechanism is expected rather than wall base flexural hinging, and vice versa.

2.3. Influence of Uncertainties on the Expected Mechanism?

The previous sections have presented a series of equations and a procedure for identifying the likely mechanism. However, the actual resistance a structure offers against formation of a certain type of mechanism is uncertain. Furthermore, the uncertainty in the resistance against some types of mechanisms may be higher than others. Consider, for example, the scenario depicted in Figure 3 where the uncertainty in the shear resistance of a wall is deemed greater than the uncertainty in the flexural resistance. The figure shows that even though a seismic assessment undertaken considering median values of resistance may indicate that a flexural mechanism is more likely, at low intensity levels (below the intersection point, IP) the likelihood of reaching the failure limit state may be dictated by the shear resistance. Consequently, the annual probability of exceeding the failure limit state in this case should be computed considering both mechanisms. Hence, for cases where there is doubt about which mechanism may form, the engineer should undertake the seismic assessment for each mechanism conditioned on the other mechanism not forming, and then later evaluate the overall probability of failure.

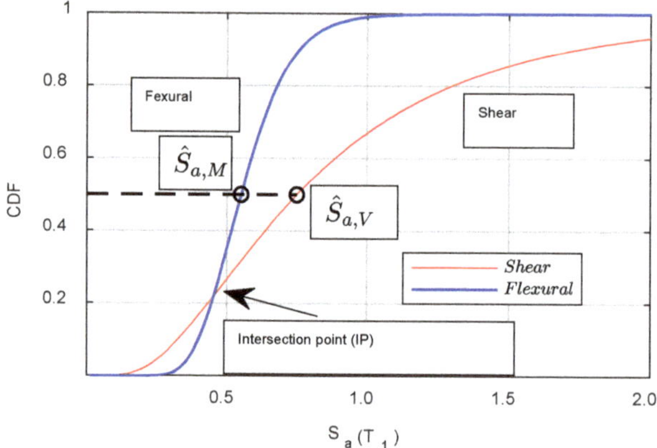

Figure 3. Shear and flexural failure mechanism probability distribution.

In order to compute the probability of exceeding the limit state for a given mechanism, however, one must first identify the median intensity capacity of each given mechanism. This in turn requires assessment of the expected displacement capacity of each mechanism for different limit-state triggers (e.g., excessive story drift demand or excessive wall curvature demand). Guidance on this will be provided in the next section.

3. Identification of Limit State Deformation Capacity

The equivalent SDOF displacement capacity of a structure, shown as Δ_{cap} (also referred to as Δ_{sys}) in Figure 1, will be a function of the deformation capacity of the structural and non-structural elements as well as the displaced shape of the structure. The displaced shape will in turn depend on the expected mechanism. The following sub-sections illustrate how the displaced shape can be computed considering different failure mechanisms.

3.1. Displacement Capacity for the Case of Flexural Hinging at Wall Base

For the case of flexural hinging at the wall base, the limit state displacement profile, $\Delta_{ls,i}$, can be found as (Priestley et al. [11], Sullivan et al. [35]):

$$\Delta_{ls,i} = \Delta_{y,i} + \Delta_{p,i}, \tag{16}$$

where $\Delta_{y,i}$ is the displacement profile at yield of the walls, given by Equation (10), and $\Delta_{p,i}$ is the allowable plastic displacement component for the limit state, computed as:

$$\Delta_{p,i} = h_i \theta_{p,min}, \tag{17}$$

where $\theta_{p,min}$ is the minimum allowable plastic hinge rotation at the base of the wall, considering structural and non-structural deformation limits. This can be found by taking the minimum of $\theta_{p,NS}$ and $\theta_{p,S}$ from below:

$$\theta_{p,NS} = \theta_c - \frac{\phi_y H}{2} \tag{18}$$

$$\theta_{p,S} = (\phi_{ls} - \phi_y) L_p, \tag{19}$$

where θ_c is the nonstructural drift limit capacity, ϕ_{ls} is the limit state curvature capacity, ϕ_y is the wall section yield curvature and L_p is the wall plastic hinge length. The plastic hinge length proposed by Priestley et al. [11] is given by:

$$L_p = kH_e + 0.1L_w + L_{sp}, \tag{20}$$

$$k = 0.2\left(\frac{f_u}{f_y} - 1\right) \le 0.08, \tag{21}$$

$$L_{sp} = 0.022 \times f_{ye} \times d_b, \tag{22}$$

where L_{sp} is strain penetration length, f_{ye} is the effective yield stress in MPa, and the other parameters have been defined, previously. However, if a different expression for the plastic hinge length were deemed more appropriate, it could be substituted into Equation (20).

3.2. Displacement Capacity for the Case of a Wall Shear Mechanism

For the case that a wall shear mechanism is expected to precede flexural hinging, the limit state displacement profile, $\Delta_{i,ls}$, can be approximated as:

$$\Delta_{ls,i} = \frac{V_b}{\omega_v . \varphi_o V_y} \Delta_{y,i}, \tag{23}$$

where V_b is the base shear capacity associated with the shear mechanism, $\Delta_{y,i}$ is the displacement profile of the wall that would be expected if flexural yielding were to occur and the terms in the denominator represent the base shear force at flexural yielding amplified to account for higher mode effects and overstrength. This expression should only be applied for cases where flexural cracking of the wall is expected prior to shear failure. If flexural cracking is not expected, the displacements will be considerably lower and could be computed using elastic analysis with gross (uncracked) wall section properties.

Another type of shear mechanism occurs when initially, flexural yielding begins to occur but subsequently, due to a loss in wall shear resistance at increasing ductility demands, a shear failure is triggered. For this case, the displacement capacity can be estimated by finding the curvature ductility demand that causes the shear strength to first drop to the shear demand level. The displacement profile is then computed for this value of curvature demand using Equations (16)–(19). This will be clarified further as part of a case study example later in this paper.

3.3. Substitute Structure Characteristics

Once the displacement profile at the limit state is identified, the equivalent SDOF displacement capacity, effective mass, and effective height can be found as per the equations below (refer to Priestley et al. [11]).

$$\Delta_{cap} = \frac{\sum m_i \Delta_i^2}{\sum m_i \Delta_i}, \tag{24}$$

$$m_e = \frac{(\sum m_i \Delta_i)^2}{\sum m_i \Delta_i^2},\tag{25}$$

$$H_e = \frac{\sum m_i \Delta_i h_i}{\sum m_i \Delta_i},\tag{26}$$

where Δ_i represents the yield displacement at story i for the first case, and the ultimate limit state displacement at story i for the second case. The substitute structure effective mass, m_e, can be computed by adopting Equation (25), but adopting different displacement profiles associated with different limit states may change the mass found for the substitute structure. Furthermore, the effective height can be achieved by adopting Equation (26) employing the displaced shape, Δ_i, story mass, m_i, and height, h_i.

By applying Equation (24) using the displacement profile at yield and also the limit state displacement profile, the substitute structure's yield displacement, Δ_y, and limit state displacement capacity, Δ_{cap}, are obtained, respectively. As such, the displacement ductility ratio, μ_Δ, is achieved by dividing the limit state displacement by the yield displacement, as illustrated Equation (27).

$$\mu_\Delta = \frac{\Delta_{cap}}{\Delta_y},\tag{27}$$

where all parameters have been defined, previously.

4. Quantification of the Limit State Median Intensity Capacity and Annual Probability of Failure

With the base shear resistance, the equivalent SDOF displacement capacity, and the ductility computed, the limit state seismic intensity capacity can be found. This is done by firstly estimating the spectral acceleration required to cause yield; this is done by dividing the yield base shear by the equivalent SDOF seismic mass. The period of vibration of the building is then computed using Equation (1) and considering the Takeda Thin hysteretic model for RC walls, an appropriate b factor is read from Table 1 (with interpolation if required). Consequently, the b factor, yield spectral acceleration, S_{ay}, and ductility factor (μ) are implemented into Equation (28) to estimate the median spectral acceleration capacity expected:

$$S_a = \left(1 + (\mu - 1)^{\frac{1}{b}}\right) S_{ay} = \left(1 + \left(\frac{\Delta_{cap}}{\Delta_y} - 1\right)^{\frac{1}{b}}\right) S_{ay}\tag{28}$$

where all parameters have been defined previously.

Note that the base shear resistance identified during the mechanism assessment process described in Section 2 is the first-order resistance and second-order P-delta effects may also need to be accounted for. P-delta effects reduce the equivalent lateral force resistance because of the additional overturning moment demands induced by geometric nonlinearity. As is illustrated in Figure 4, the lateral forces represented as F deform the structure by the amount of Δ. This lateral movement, however, causes the gravity loads to induce some extra bending moment demand and hence the equivalent lateral force resistance reduces as illustrated in Figure 4 and Equation (29) (refer Priestley et al. [11] and Sullivan et al. [13]).

$$V_{b-P\Delta} = V_b - \frac{P\Delta}{H_e} = V_b - \frac{m_e g . \Delta_{cap}}{H_e},\tag{29}$$

$$\theta_{P\Delta} = \frac{P\Delta}{V_b H_e},\tag{30}$$

where $V_{b-P\Delta}$ is the base shear capacity reduced to account for P-delta effects, P represents the effective gravity loads, Δ is the substitute structure lateral displacement, H_e is the effective height.

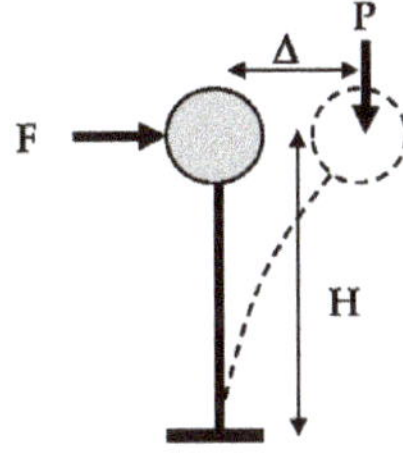
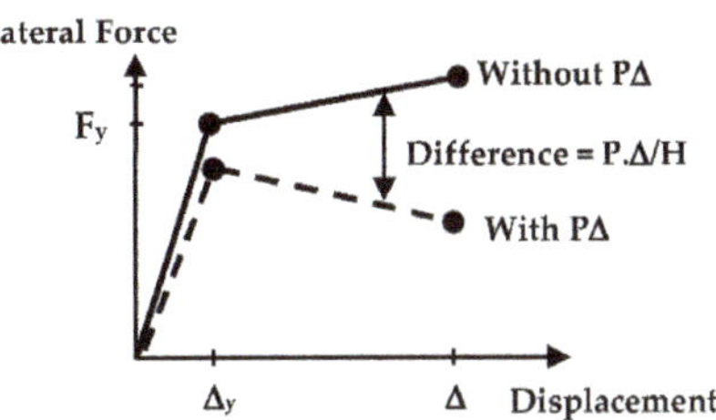

(a) SDOF system subject to lateral force F and axial load P.

(b) Force-displacement response of SDOF system with and without P-Δ.

Figure 4. Influence of second-order P-delta effect on the lateral force-displacement response of a SDOF system (Adopted with permission from [13]. Copyright 2013 IUSS press; F is the total lateral force equivalent to base shear before considering geometric nonlinearity).

As part of an assessment of the vulnerability of a structure to second-order *P*-delta effects, the stability coefficient ($\theta_{P\Delta}$) should be computed as shown in Equation (30). This coefficient measures the ratio between the P-delta induced overturning moments and the overturning resistance. Note that the stability factor ($\theta_{P\Delta}$) should be less than 0.3 in order to limit the likelihood of collapse due to dynamic instability, or the value of Δ_{cap} reduced such that the limit is satisfied.

Once the median intensity required to exceed the limit state has been identified, the annual probability of exceeding the limit state can be computed, as will be explained in the next section.

5. Quantifying the Annual Probability of Exceeding the Assessment Limit State

To compute the annual probability of exceeding a limit state, the procedure described in Section 1.1 can be followed. This requires information on the mean annual frequency of exceeding different ground shaking intensity levels, and the corresponding hazard fit curve coefficients, as illustrated in Equation (3). Furthermore, estimates of ground motion variability and capacity uncertainty are necessary as illustrated in Equation (5). Consequently, the annual probability of limit state exceedance is computed by adopting Equation (2) together with the estimated dispersion and hazard.

The risk of performance failure for systems with multiple failure mechanisms should be computed by considering all relevant mechanisms. For cases in which the failure mechanisms are independent and in series, Lupoi et al. [36] provide Equation (31) to compute the likelihood of failure considering more than one possible mechanism. Specifically, the equation considers the likelihood that element *i* may fail in each considered failure mechanism *m*. Hence, one could consider multiple walls (and other structural and non-structural elements) as well as multiple possible mechanisms.

$$P_{f,k} = \Pr\left[\bigcup_{i=1}^{m} C_i \leq D_{ik}\right] = 1 - \prod_{i=1}^{m}\left(1 - P_{f,ik}\right) \tag{31}$$

where C_i is the *i*th element's capacity and D_{ik} is the *i*th element's demand because of *k*th accelerograms at *m*th mechanism; $P_{f,ik}$ is the probability of *i*th element failure because of *k*th accelerograms at *m*th mechanism. The first term of Equation (31) illustrates the probability of system failure by virtue of activation of *m* failure mechanisms in general, which has been simplified into the second term as per the assumptions above.

6. Gauging the Accuracy of Probabilistic Displacement-Based Assessment

To gauge the accuracy and limitations of the proposed approach, a series of case-study buildings of 4-, 8-, and 12-storys, are assessed in this section. The layout of the three case study buildings is illustrated in Figure 5.

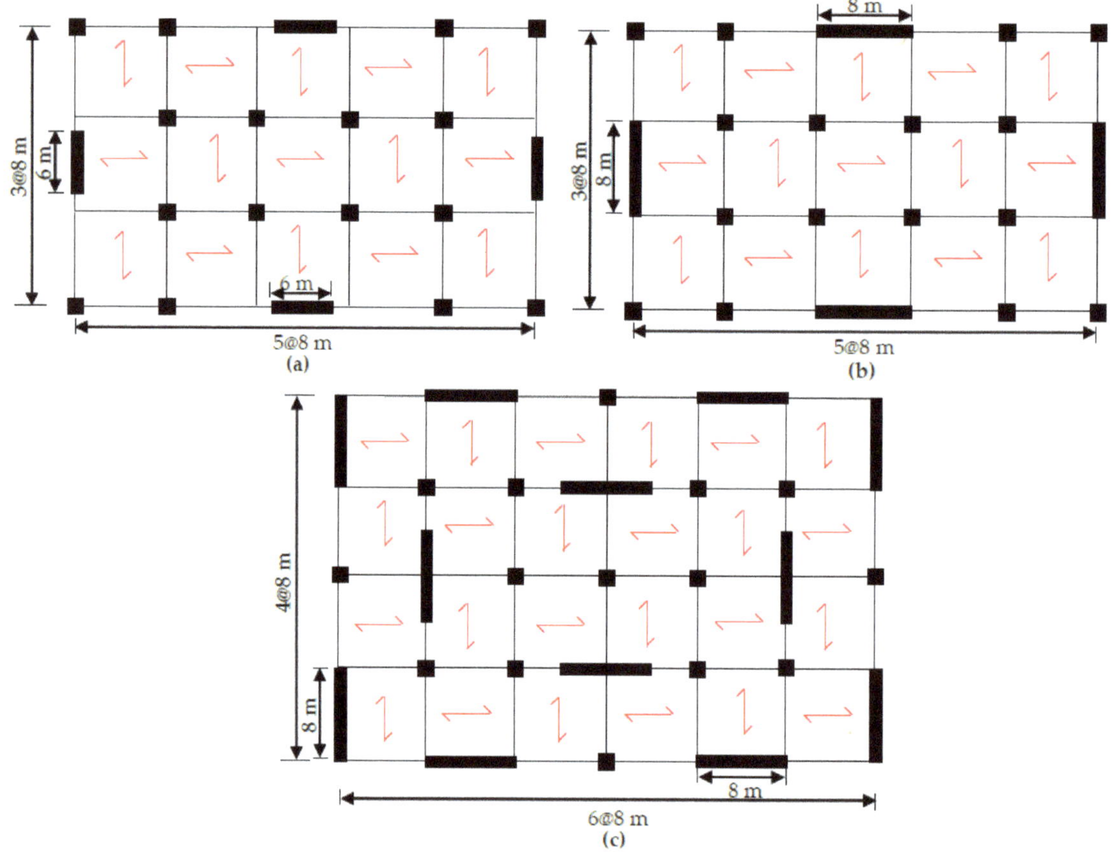

Figure 5. Case study buildings: (**a**) 4-story; (**b**) 8-story; (**c**) 12-story reinforced concrete wall buildings' layout.

The buildings have not been designed and are instead idealized layouts of buildings, not representative of any particular buildings but with dimensions and reinforcement quantities that are deemed realistic for regular New Zealand RC wall buildings. The 4-story case study building possesses two 6 m walls in each direction, and two 8 m walls are located in each direction of the 8-story, with six 8 m walls located in each principle direction of the 12-story case study building. The thickness of the walls is 0.3 m for all case studies, and the story height is 3.6 m except for the first story, which is 4.5 m.

The wall axial load ratio ($N/f'_c A_g$) is 2.3%, 3.5%, 5.3%, and the longitudinal reinforcement ratio is 0.85%, 1.1%, and 1.6% for the 4-, 8-, and 12-story walls, respectively. However, in order to investigate possible mixed shear-flexure mechanisms, the transversal reinforcements spacing is taken as 170 mm for the 4-story, and 200 mm for the 8- and 12-story buildings; and the transversal rebar diameter is 10 mm for all case study buildings. Furthermore, the characteristic concrete strength is 30 MPa, and the reinforcement's effective yield and ultimate strength is taken to be 500 MPa and 650 MPa, respectively. The median seismic mass is taken as 602 t for the 4- and 8-story, and 722 t for the 12-story building at each level (noting that variations in mass are considered as part of the probabilistic seismic assessment). The median base shear at yield of the reinforced concrete walls is 3438 kN, 3664 kN, and 13,834 kN for the 4-, 8-, and 12-story buildings, respectively. These base shear resistances correspond to design strength coefficients (lateral yield strength divided by weight) that are again typical of existing buildings in New Zealand, ranging from 0.10 to 0.19.

6.1. Rigorous Probabilistic Assessment of the Multi Story RC Wall Buildings

Multi-stripe analyses were conducted using lumped plasticity models developed in Ruaumoko [37]. The RC walls were modeled using their cracked section properties of I_{eff} = 0.84 m^4, A = 1.8 m^2, A_v = 0.38 m^2 for the 4-story, I_{eff} = 2.24 m^4, A = 2.4 m^2, A_v = 0.42 m^2 for the 8-story, and I_{eff} = 4.12 m^4, A = 2.4 m^2, A_v = 0.64 m^2, for the12-story walls. The wall flexural nonlinear behavior was modeled by employing Giberson beam elements with the Takeda hysteresis rule indicating alpha and beta equal to 0.5 and 0.0 [11], respectively. The foundations were assumed rigid for all case study buildings. The tangent-stiffness proportional Rayleigh damping model was used with 5% damping specified at the fundamental mode and the other one with more than 90% mass contribution. The fundamental mode was found to be 1.0 s for the 4-story building and 2.0 s for the 8- and 12-story case study buildings (adopting cracked section properties). The ground motions employed for MSA were selected to be hazard consistent by [38] for a conditioning period of 1.0 s for the 4-story case study and 2.0 s for the 8- and 12-story case studies for nine different intensity levels (stripes) at a soil-type C site in Wellington [38], following the generalized conditional intensity measure procedure detailed in [39].

The variability caused by epistemic uncertainty was introduced to the finite element models by assuming the capacity parameters are log normally distributed as illustrated in Figure 6, with the median and dispersion values listed in Table 3. It can be seen that the random variables include the wall yield strength, M_y, the wall effective (i.e., cracked) section second moment of inertia, I_{eff}, the wall yield curvature, ϕ_y, and the ultimate curvature capacity of the walls, ϕ_{ult}, as well as the seismic mass (with dispersion of 0.1) and the modeling damping coefficient (with dispersion of 0.6). These variables were selected noting that O'Reilly et al. [40] and Gokkaya et al. [41] had noted they tended to have the most significant impact on the seismic assessment results of existing RC buildings. The sampling procedure followed by [40] was used herein. As such, the available random function in MATLAB (2017b) [42] was utilized to generate 250 samples from each structural parameter distribution, as illustrated in Figure 6. Hence, the randomly generated sample structures were modeled in Ruaumoko [37] and exposed to 40 ground motions selected at each stripe leading to 10,000 nonlinear analyses.

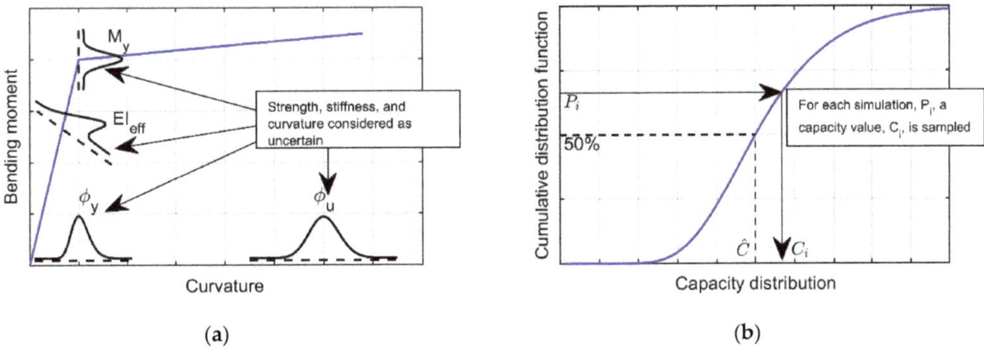

Figure 6. Probabilistic modeling approach (**a**) overview of variables included in the moment-curvature capacity curve and (**b**) sampling process for each capacity parameter.

Table 3. Uncertain modeling parameters considered for three case study buildings.

Case Study *	M_y (kN·m)	β	I_{eff} (m^4)	β	ϕ_y(m^{-1})	β	ϕ_{ult}(m^{-1})	β
4-story	20,820	0.3	0.84	0.3	8.33×10^{-4}	0.3	3.5×10^{-3}	0.3
8-story	41,677	0.3	2.24	0.3	6.25×10^{-5}	0.3	2.62×10^{-3}	0.3
12-story	76,558	0.3	4.12	0.3	6.25×10^{-5}	0.3	2.62×10^{-3}	0.3

* Note that in addition to the parameters listed above, dispersions of 0.6 and 0.1 were considered for damping coefficient and seismic mass, respectively [40].

The displacement ductility demand found for each structure and ground motion were utilized to compute the median shear capacity (V_{prob}) following the algorithm developed by Orumiyehei and Sullivan [43] and NZSEE [3]. Shear failures were detected by comparing the randomly generated shear capacities and shear demand obtained for each ground motion. The probability of failing in shear at an intensity level (stripe) was achieved after dividing the number of failures by the number of randomly generated shear capacities exposed to the selected ground motions. A fragility curve was fit using the maximum likelihood technique described by Baker [44].

The other failure mechanism that was deemed possible for these case study buildings is longitudinal reinforcement buckling as per NZSEE [3] guidelines. Considering the transversal reinforcement spacing and the longitudinal rebar diameter, the curvature ductility capacities associated with rebar buckling was found to be 4.2 for the three case study buildings. Using this curvature as the median curvature ductility capacity and a dispersion of 0.3, 1000 capacity simulations were randomly generated. Consequently, the curvature demand obtained for each ground motion at each stripe was compared with the curvature capacity samples. The number of failures found at each stripe provided information on the curvature failure probability; accordingly, a fragility curve was fit adopting the maximum likelihood technique.

Seismic loss is highly correlated with displacement demands due to the large number of nonstructural and structural elements that experience damage by increasing deformation demands. As such, two drift limits of 1% and 2% were selected as hypothetical median displacement capacity limits for nonstructural and structural elements. Furthermore, a dispersion value of 0.3 is assumed as the variability associated with the displacement capacity of the aforementioned limit states. Similar to previous mechanisms, the aforementioned median and dispersion values are employed to generate drift capacities randomly, which are compared with drift demands imposed by each ground motion. Consequently, the median intensity and dispersion associated with failure of each limit state were obtained following the same procedure explained before.

The results of the multi-stripe analysis for the 4-story case study building are presented in Figure 7. The distribution of peak story drift demand at each stripe is illustrated on the left side of Figure 7, and the chances that a given mechanism may fail at a given intensity in addition to the best fit curves corresponding to the 4-story failure mechanisms are presented on the right side of Figure 7. The fragility curves for the 8- and 12-story case study buildings are presented in Figures 8 and 9, respectively. The median spectral acceleration and dispersion observed for each mechanism and building are reported in Table 4.

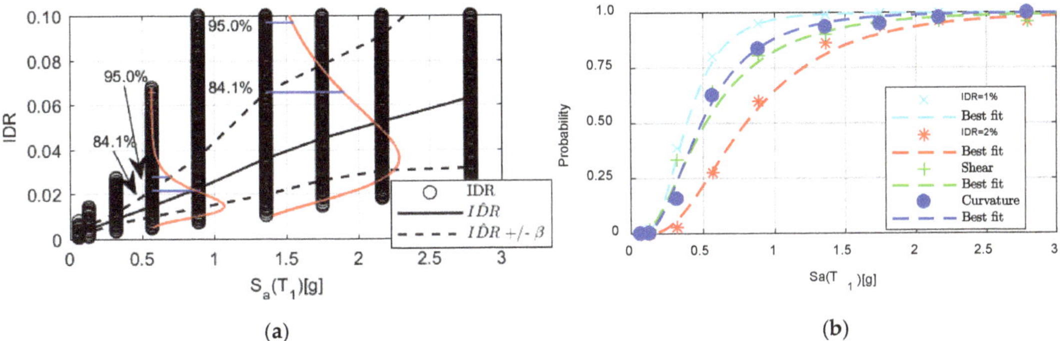

Figure 7. Multi-stripe analysis results for the 4-story case study building; (**a**) inter-story drift ratio demands obtained at each intensity level; (**b**) probability of limit state exceedance and associated fragility curve.

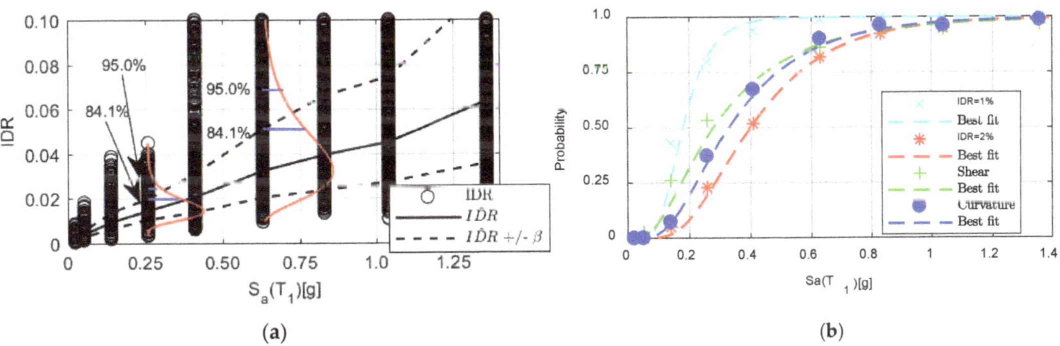

Figure 8. Multi-stripe analysis results for the 8-story case study building; (**a**) inter-story drift ratio demands obtained at each intensity level; (**b**) probability of limit state exceedance and associated fragility curve.

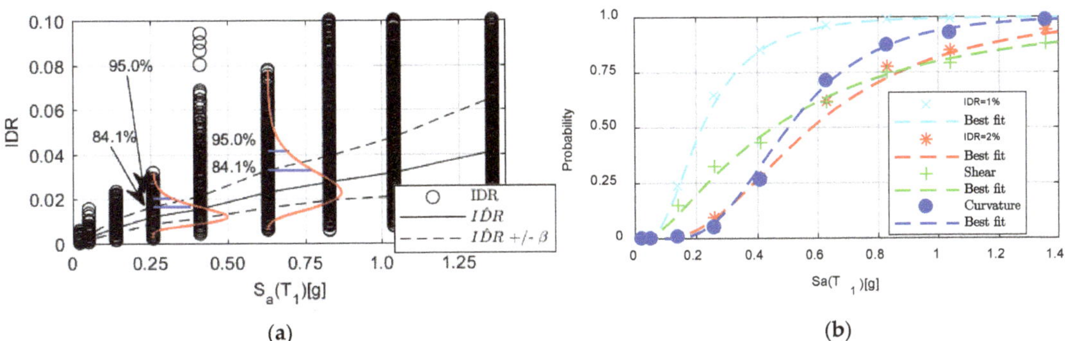

Figure 9. Multi-stripe analysis results for the 12-story case study building; (**a**) inter-story drift ratio demands obtained at each in-tensity level; (**b**) probability of limit state exceedance and associated fragility curve.

Table 4. Median intensity and dispersion for different failure limit states obtained via rigorous MSA and Mont Carlo simulations.

Limit State	4-Story		8-Story		12-Story	
	$\hat{S}_a$[g]	β	$\hat{S}_a$[g]	β	$\hat{S}_a$[g]	β
IDR = 0.01	0.39	0.50	0.18	0.40	0.22	0.60
IDR = 0.02	0.80	0.45	0.40	0.50	0.58	0.70
Shear failure	0.52	0.70	0.28	0.70	0.45	0.95
Curvature failure	0.49	0.55	0.33	0.58	0.50	0.45

The annual probability of limit state exceedance can be achieved by combining the corresponding fragility and hazard curves. The fragility curve indicates how likely the performance failure is at a given intensity measure. However, the hazard carve provides information on the annual frequency with which a given intensity measure is exceeded. As such, combining these two curves yields the annual probability of exceeding a specific limit state. This procedure has been followed for each of the mechanisms for which the structures was deemed vulnerable; the annual probability of performance failure was then computed conditioning on each failure limit state that may develop, so that the likelihood of failure for different limit states could be obtained, as presented in Table 5.

Table 5. Annual probability of limit state exceedance obtained for the three case study buildings through rigorous analysis.

Mechanism	APOE [*] 4-Story	APOE 8-Story	APOE 12-Story
IDR1 = 0.01	38×10^{-4}	35×10^{-4}	31×10^{-4}
IDR2 = 0.02	15×10^{-4}	13×10^{-4}	8×10^{-4}
Shear	32×10^{-4}	25×10^{-4}	18×10^{-4}
Curvature	31×10^{-4}	18×10^{-4}	9×10^{-4}

[*] APOE: annual probability of exceedance.

6.2. Simplified Probabilistic Assessment of the Multi Story RC Wall Buildings

The accuracy of the newly proposed assessment approach is gauged here by firstly comparing the median intensities assessed to cause the structure to violate a given limit state's displacement capacity. Secondly, the annual probability of exceeding a given limit state capacity obtained from the simplified method is compared with that found from rigorous analysis.

To estimate the median spectral accelerations associated with exceeding the first two limit states ($I\hat{D}R_1 = 0.01 \ and \ I\hat{D}R_2 = 0.02$), the steps described in Sections 2–4 are followed. For that purpose, the yield curvature is first approximated by applying Equation (11) using the reinforcement yield strain and wall length. Consequently, this curvature can be implemented in Equation (10) to estimate the yield displacement profile. However, because Equation (10) accounts for tension shift effects and a non-uniform distribution of cracking, which does not match the assumptions made in the modeling described in Section 6.1, the yield displacement profile for these case study buildings was found using the approach of Pennucci et al. [45], reported in Appendix A. Following the displacement based assessment approach [11] the equivalent single degree of freedom yield displacement is then obtained. Furthermore, the displacement demand over the height of the building at a given limit state is found as illustrated in Appendix A. This displacement profile can be employed to reach the equivalent SDOF displacement capacity for the target limit state. Subsequently, the ductility ratio is computed by dividing the equivalent SDOF system's displacement capacity at the target limit state by the yield displacement. Using the ductility ratio, Equation (28) estimates the median spectral acceleration capacity associated with the limit state. The median spectral accelerations that were assessed to cause the first two limit states to be exceeded are reported in Table 4.

The median spectral acceleration associated with exceeding the shear capacity can be estimated by comparing the flexural capacity curve with the shear capacity found using the modified UCSD model [3]. Shear capacity, however, was calculated as a function of displacement ductility, as illustrated in Appendix A [46]. Furthermore, the curvature capacity was found by, firstly, assessing the strain associated with buckling of vertical reinforcement, as illustrated by Equation (32) [3], and subsequently, the curvature capacity can be computed by implementing the achieved strain in Equation (33) [3]. The median spectral acceleration associated with exceeding the curvature capacity for the three case study buildings are compared with those obtained from MSA and presented in Table 6. The results shown in Table 6 provide evidence that the displacement-based assessment method can effectively assess the intensity at which key limit states are exceeded. The details regarding the above calculation are presented in Appendix A.

$$\varepsilon_p^* = \frac{11 - \frac{5}{4}(s/d_b)}{100} \tag{32}$$

$$\phi_{cap} = \frac{\varepsilon_p^*}{Yl_w} \tag{33}$$

where s is the transversal reinforcement spacing, d_b is the longitudinal reinforcement diameter, ε_p^* is the strain that triggers the vertical reinforcement buckling, ϕ_{cap} is the curvature capacity, Yl_w is the length of wall after ignoring cover thickness.

Table 6. Median intensity associated with different failure mechanisms found through the simplified approach and multi-stripe analysis (MSA).

Limit States	4-Story, $S_a(T_1)$ [g]		8-Story, $S_a(T_1)$ [g]		12-Story, $S_a(T_1)$ [g]	
	Simplified	MSA	Simplified	MSA	Simplified	MSA
IDR = 0.01	0.39	0.39	0.20	0.18	0.30	0.22
IDR = 0.02	0.78	0.80	0.39	0.40	0.57	0.58
Shear	0.58	0.52	0.34	0.28	0.50	0.45
Curvature	0.43	0.49	0.34	0.33	0.46	0.50

The annual probability of exceeding each limit state was also computed through the simplified approach by adopting the closed-form solution [16], as illustrated in Equations (2)–(5). Note that the annual probability of performance failure has been calculated through the simplified approach by adopting assumed dispersion (β = 0.45 for 1% and 2% drift capacity limit state, β = 0.5 for flexural, and β = 0.75 for shear failure) and compared against those achieved through rigorous analysis as presented in Table 7. Furthermore, to highlight the effect of uncertainty estimation in simplified approach, the likelihood of limit state failure has been also computed using the numerically found dispersion as illustrated in Table 8. It can be seen that the simplified approach is able to provide good estimates of the annual probability of exceeding key limit states but that the accuracy is quite dependent on the assumed dispersion.

Table 7. Annual probability of exceeding different limit states for the case study buildings obtained through rigorous assessment and via the simplified approach with assumed values of total dispersion.

Limit States	APOE * 4-Story		APOE 8-Story		APOE 12-Story	
	Simplified β_{Assumed}	Rigorous β_{Rigorous}	Simplified β_{Assumed}	Rigorous β_{Rigorous}	Simplified β_{Assumed}	Rigorous β_{Rigorous}
IDR1	3.94×10^{-3}	3.80×10^{-3}	3.23×10^{-3}	3.50×10^{-3}	2.00×10^{-3}	3.10×10^{-3}
IDR2	1.50×10^{-3}	1.50×10^{-3}	1.34×10^{-3}	1.30×10^{-3}	7.90×10^{-4}	8.40×10^{-4}
Shear	3.13×10^{-3}	3.20×10^{-3}	2.74×10^{-3}	2.50×10^{-3}	1.29×10^{-3}	1.77×10^{-3}
Curvature	3.60×10^{-3}	3.10×10^{-3}	1.75×10^{-3}	1.80×10^{-3}	1.11×10^{-3}	9.00×10^{-4}

* APOE: annual probability of exceedance.

Table 8. Annual probability of exceeding limit states for the case study buildings obtained through rigorous multi-stripe analyses and via the simplified approach using rigorous (numerically found) values of total dispersion.

Limit States	APOE * 4-Story		APOE 8-Story		APOE 12-Story	
	Simplified $\beta_{Rigorous}$	Rigorous $\beta_{Rigorous}$	Simplified $\beta_{Rigorous}$	Rigorous $\beta_{Rigorous}$	Simplified $\beta_{Rigorous}$	Rigorous $\beta_{Rigorous}$
IDR1	4.10×10^{-3}	3.80×10^{-3}	3.13×10^{-3}	3.50×10^{-3}	2.16×10^{-3}	3.10×10^{-3}
IDR2	1.50×10^{-3}	1.50×10^{-3}	1.39×10^{-3}	1.30×10^{-3}	1.02×10^{-3}	8.40×10^{-4}
Shear	2.95×10^{-3}	3.20×10^{-3}	2.60×10^{-3}	2.50×10^{-3}	1.68×10^{-3}	1.77×10^{-3}
Curvature	3.80×10^{-3}	3.10×10^{-3}	1.87×10^{-3}	1.80×10^{-3}	1.06×10^{-3}	9.00×10^{-4}

* APOE: annual probability of exceedance.

7. Discussion

The process followed in this work has been seen to enable estimates of seismic risk (the annual probability of exceeding key limit states) to be computed in a somewhat simplified fashion. A question that could be asked, however, is "Would such information on seismic risk change decisions?" To reflect on this question, consider the results for an 8-story building reaching the limit state because of a 2% drift limit as opposed to the results for the 12-story building reaching the flexural–shear failure limit. These case study buildings have the same period (2s) and, hence, looking at the median intensity capacity results (from Table 6 we see a capacity of S_a = 0.40 g for the 8-story building and a capacity of S_a = 0.45 g for the 12 story building), one would anticipate that the seismic risk for the 8-story building is higher than that of the 12-story building. However, from Table 8, we see that the 12-story building actually has a higher annual probability of exceeding the shear limit state than the 8-story building has of exceeding the 2% drift limit (the annual probability of exceeding the limit state is 0.00130 for the 8-story building and is 0.00177 for the 12-story building) and, thus, should be more of a priority for seismic retrofit. This highlights the fact that computing the seismic risk could indeed impact decision making in practice.

The results presented in Table 8 also provide evidence of clear differences between 4-story and 12-story performances. It is seen that the 12-story case study building has less chance of experiencing shear failure despite similar transversal reinforcement detailing. Furthermore, assuming identical curvature ductility capacity for both the 4-story and 12-story case studies, the chances for vertical reinforcement buckling for the 4-story walls is almost three times larger than that found for the 12-story walls. However, comparing the chances of curvature failure with that of the exceeding 2% drift capacity, it seems that the ratio of these chances is about two for the 4-story building. This is almost two times the ratio achieved, comparing the failure likelihoods for the same limit states chances for the 12-story case study building. These odds reflect the tendency for walls with high-aspect ratio (height divided by wall length) to be more sensitive to drift demands than curvature demands.

The results for the 8-story case study building found from rigorous analysis align with the transition in behavior expected from short 4-story structures towards the taller 12-story structures. The simplified method has successfully predicted the failure chances for different limit states with acceptable accuracy. As such, the approach is deemed capable of differentiating between failure mechanisms and structural behavior.

The total annual probability of performance failure could be achieved by accounting for multiple failure mechanisms. As reported earlier, Lupoi et al. [36] proposed Equation (29) which can be adopted for systems with multiple failure mechanisms. Furthermore, the change in the annual probability of limit state exceedance caused by retrofit decisions that suppress different failure mechanisms could be investigated. In line with this, Table 9 reports the annual probability of limit state exceedance for each failure mechanism separately as well as for all four simultaneously. For instance, the annual probability of exceeding a drift capacity of 1% and 2% for the 4-story case study is 4.10×10^{-3}, 1.50×10^{-3}, respectively. However, it changes to 2.95×10^{-3}, 3.35×10^{-3} for curvature and shear failure, respectively. As such, the system performance failure assuming all failure mechanisms are independent and may occur simultaneously can be computed as follows.

$$P_{f,k} = 1 - \prod_{i=1}^{m}\left(1 - P_{f,ik}\right)$$

$$P_{4f,k} = 1 - \left(1 - 4.10 \times 10^{-3}\right)\left(1 - 1.50 \times 10^{-3}\right)\left(1 - 2.95 \times 10^{-3}\right)\left(1 - 3.35 \times 10^{-3}\right) = 11.8 \times 10^{-3}$$

Table 9. Annual probability of exceedance limit state accounting for three failure mechanisms before and after the retrofitting process.

Building	LS1	LS2	LS3	LS4	Simultaneous Failures and Retrofit			
					Retrofit None	Retrofit LS[1]	Retrofit LS(1,2)	Retrofit LS(1,2,3)
4-story	4.10×10^{-3}	1.5×10^{-3}	2.95×10^{-3}	3.35×10^{-3}	11.8×10^{-3}	7.78×10^{-3}	6.29×10^{-3}	3.35×10^{-3}
8-story	3.13×10^{-3}	1.39×10^{-3}	2.60×10^{-3}	1.87×10^{-3}	8.97×10^{-3}	5.85×10^{-3}	4.47×10^{-3}	1.87×10^{-3}
12-story	2.16×10^{-3}	1.02×10^{-3}	1.68×10^{-3}	1.06×10^{-3}	5.91×10^{-3}	3.76×10^{-3}	2.75×10^{-3}	1.06×10^{-3}

Assuming a (hypothetical) retrofitting process that suppresses first failure mechanism, then the system annual probability of failure would change according to the calculation is following.

$$P_{3f,k} = 1 - \left(1 - 1.50 \times 10^{-3}\right)\left(1 - 2.95 \times 10^{-3}\right)\left(1 - 3.35 \times 10^{-3}\right) = 7.78 \times 10^{-3}$$

Furthermore, the system annual probability of performance failure caused by additional retrofitting to the second and third failure mechanisms would be as follows.

$$P_{2f,k} = 1 - \left(1 - 2.95 \times 10^{-3}\right)\left(1 - 3.35 \times 10^{-3}\right) = 6.29 \times 10^{-3}$$

$$P_{1f,k} = 1 - \left(1 - 3.35 \times 10^{-3}\right) = 3.35 \times 10^{-3}$$

This calculation has been iterated for the other two case study buildings and the results are presented in Table 9. Note that this brief exercise in relation to retrofit has been presented solely with the purpose of highlighting the potential means of considering different failure mechanisms as part of a simplified probabilistic seismic assessment and retrofit procedure. In reality, the engineer would need to check whether any retrofit measures could impact the structural response (which might in turn affect the likelihood of other mechanisms forming and limit states being reached), and should also check that the retrofit measures are able to effectively reduce the probability of failure of the "strengthened" failure mechanism to zero or continue to account for this within the probabilistic assessment.

8. Conclusions

The probabilistic displacement-based assessment approach put forward for single degree of freedom systems by Orumiyehei and Sullivan [12] has been extended to reinforced concrete wall buildings in this work. The resultant simplified assessment approach can be used in combination with a set of equations [16] to assess the building's seismic risk, in terms of the annual probability of exceeding key limit states. The newly proposed approach is trialed via application to 4-, 8-, and 12-story RC wall buildings.

Different possible failure criteria are considered for the RC wall buildings, considering two possible drift limits (that could trigger non-structural or secondary structural damage), and shear and curvature failure are presumed as two commonly observed failure mechanisms in RC wall buildings. The case study buildings are assessed by conducting multi-stripe analysis and Monte Carlo simulations to generate results as a benchmark to gauge the accuracy of simplified method. The results show that the proposed approach can practically estimate the median intensity that can trigger damage to a given group of elements, as well as the annual probability of exceeding key limit states. Furthermore, this information could aid with retrofit decision making.

As the skills and tools available to the engineering profession increase, one could argue that practitioners would more accurately quantify the likelihood of exceeding limit states via multi-stripe analyses with consideration of appropriate random variables. However, considering the current assessment guidelines in New Zealand [3] that utilize displacement-based assessment, it would appear that the extension to the DBA method proposed in this work represents a valuable contribution to the state-of-the-art for practice-oriented methods. One of the main limitations with the simplified method is that it requires good estimates of the dispersion in the capacity. Future research should seek to provide more information on the dispersion for different systems and limit state failure types. Furthermore, the method has not been developed or validated for asymmetric buildings and this would be another area for future research.

Author Contributions: Conceptualization, T.J.S. and A.O.; methodology, T.J.S. and A.O; software, T.J.S. and A.O.; validation T.J.S. and A.O.; formal analysis, T.J.S. and A.O.; investigation, T.J.S. and A.O.; resources, T.J.S.; data curation, T.J.S. and A.O.; writing—original draft preparation, T.J.S. and A.O.; writing—review and editing, T.J.S. and A.O; visualization, T.J.S. and A.O.; supervision, T.J.S.; project administration, T.J.S.; funding acquisition, T.J.S. All authors have read and agreed to the published version of the manuscript.

Funding: This project was partially supported by QuakeCoRE, a New Zealand Tertiary Education Commission-funded Centre. This is QuakeCoRE publication number 0689.

Conflicts of Interest: The authors declare no conflict of interest.

Appendix A

This appendix provides detailed information on the steps taken to assess the structural behavior of the case study buildings. Hence, the procedure that has been followed to find the median spectral acceleration associated with exceeding 1% and 2% as the maximum drift ratio capacities, in addition to that of exceeding the shear capacity and curvature ductility capacity are demonstrated for the 4-story case study building, illustrated in Figure 5 of Section 6.

The first limit state intensity can be estimated by applying Equation (28) illustrated in Section 4. For that purpose, the displacement ductility should be estimated as shown in Equation (27). Hence, the yield curvature is calculated using Equation (11) explained in Section 2.1, and repeated below for illustration.

$$\phi_y = \frac{e\varepsilon_y}{D},$$

$$D = L_w = 6.0\ m, \quad e = 2, \quad \varepsilon_y = \frac{500}{200,000} = 0.0025,$$

$$\phi_y = 2 \times \frac{0.0025}{6} = 8.3 \times 10^{-4} \left[\frac{1}{m}\right]$$

After finding the yield curvature, it is implemented in Equation (A1) [45] and the yield displacement at different story heights is estimated, as follows.

$$\Delta_{y,x}(x) = \frac{3\phi_y}{H^3} \left[\frac{x^5}{120} - \frac{H^2}{12} x^3 + \frac{H^3}{6} x^2 \right] \tag{A1}$$

$$\phi_y = 8.3 \times 10^{-4} \left[\frac{1}{m}\right], \; H = 15.3[m], \; and \; x = 4.5, \; 8.1, \; 11.7, \; 15.3[m]$$

$$\Delta_{y,i} = 7.20 \times 10^{-3}[m], \quad 20.3 \times 10^{-3}[m], \quad 36.5 \times 10^{-3}[m], \quad 53.6 \times 10^{-3}[m]$$

The resulting yield displacement profile and seismic mass are inserted into Equation (24) explained in Section 3.3, and the SDOF equivalent yield displacement is found to be 0.04 m. Furthermore, the associated limit state displacement demand at different levels can be estimated using Equation (A2), and consequently, adopted in Equation (24). Following this approach, the SDOF equivalent limit state displacement is found to be $8.1 - 10^{-2}$ and displacement ductility of 2.0, as illustrated below.

$$m_{1,2,3,4} = 301 \; ton; \quad \delta_{SDOF,y} = \frac{\sum m_i \Delta_{y,i}^2}{\sum m_i \Delta_{y,i}} = 4 \times 10^{-2}[m]$$

$$\delta_{LS,hi} = \Delta_{y,i} + \left(\theta_{LS} - \frac{\phi_y H}{2}\right) h_i \tag{A2}$$

$$\Delta_{IDR=0.01,i} = 2.35 \times 10^{-2}[m]; 4.97 \times 10^{-2}[m]; 7.89 \times 10^{-2}[m]; 10.91 \times 10^{-2}[m],$$
$$\delta_{SDOF,IDR=0.01} = 8.1 \times 10^{-2}[m]$$
$$\mu_\Delta = \frac{\delta_{LS}}{\delta_y} = \frac{8.1 \times 10^{-2}}{4.0 \times 10^{-2}} = 2.0$$

At this stage, knowing the yield spectral acceleration to be equal 0.195 g, the median intensity associated with exceeding 1.0% is found to be 0.39 g (as presented in Table 6) adopting Equation (28) (from Section 4) and employing a displacement ductility and b factor listed in Table 1, as illustrated below.

$$\frac{S_a}{S_{ay}} = 1 + (\mu - 1)^{\frac{1}{b}} = 1 + (2 - 1)^{\frac{1}{1.23}} = 2$$
$$S_a = 2 \times S_{ay} = 2 \times 0.195 = 0.39 \; g$$

Following a similar approach, the displacement ductility associated with the 2% drift ratio as the second limit state is found to be 4.94. Furthermore, the median intensity is found to be 0.78, as presented in Table 6 and shown below.

$$\frac{S_a}{S_{ay}} = 1 + (\mu - 1)^{\frac{1}{b}} = 1 + (4.94 - 1)^{\frac{1}{1.23}} = 4$$
$$S_a = 4 \times S_{ay} = 4 \times 0.195 = 0.78 \; g$$

Shear failure is the third mechanism that has been investigated; to assess the displacement capacity of the shear mechanism it is assumed that wall has been provided with enough curvature ductility capacity. As such, the shear capacity is calculated using the modified UCSD shear model adopted for RC wall structures [3]. After calculating shear capacity as a function of displacement ductility, the flexural capacity curve (also referred to as a pushover curve) is estimated. Furthermore, the achieved shear demand is multiplied by higher modes effect factor. The shear capacity and demand are presented in terms of acceleration in Figure A1. The displacement at which shear failure occurs is found as the intersection point of two aforementioned curves, as illustrated in Figure A1. Finally, the displacement associated with the shear failure is divided by SDOF equivalent yield displacement, and the displacement ductility found to be 3.2, which implemented in Equation (28) and, consequently, the associated spectral acceleration is achieved equal to 0.58 g (as presented in Table 6), as demonstrated below.

$$\frac{S_a}{S_{ay}} = 1 + (\mu - 1)^{\frac{1}{b}} = 1 + (3.2 - 1)^{\frac{1}{1.23}} = 2.9$$

$$S_a = 0.58 \; g$$

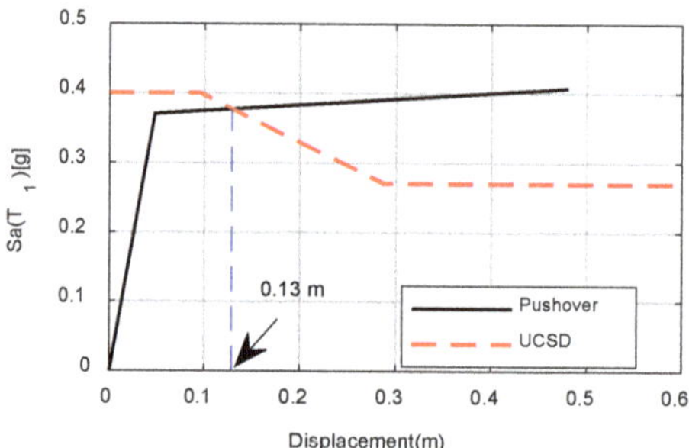

Figure A1. The modified UCSD shear capacity calculated for 4-story case study building.

Lack of sufficient restraint provided by transversal reinforcement ($s/d_b > 6$), cyclic loading, and strains larger than yield are among the parameters that limit the curvature capacity. The curvature capacity can be computed as illustrated in Equation (33) [3]. Indeed, the curvature capacity is a function of the maximum strain experienced by vertical rebar that is calculated as illustrated in Equation (32). In the case of the 4-story RC wall, transversal reinforcement spacing (s) and longitudinal rebar diameter (d_b) are assumed equal to (approximately) 170 and 24 mm, respectively. Hence, the curvature capacity is computed as follows.

$$\phi_{cap} = \frac{\varepsilon_p^*}{Yl_w},$$

$$\varepsilon_p^* = \frac{11 - \frac{5}{4}(s/d_b)}{100},$$

$$\varepsilon_p^* = \frac{11 - \frac{5}{4}(170/24)}{100} = 2.1 \times 10^{-2},$$

$$\phi_{cap} = \frac{2.1 \times 10^{-2}}{5.9} = 3.5 \times 10^{-3} \left[\frac{1}{m}\right]$$

Furthermore, the curvature ductility found to be about 4.2 assuming the yield curvature equal to 8.3×10^{-4}. As such, the median intensity associated with the rebar buckling mechanism can be computed following the displacement based design approach [11], as demonstrated below.

$$\phi_{total} - \phi_y = 3.5 \times 10^{-3} - 8.3 \times 10^{-4} = 2.67 \times 10^{-3} \left[\frac{1}{m}\right]$$

Furthermore, the plastic hinge length, strain penetration length, and SDOF equivalent yield displacement are found to be equal 1.51 m, 0.26 m, and 0.04 m, respectively. Consequently, the plastic hinge rotation, SDOF equivalent plastic displacement, and total displacement are calculated as follows.

$$\theta_p = L_p \times \phi_p = 1.51 \times 2.67 \times 10^{-3} = 4.03 \times 10^{-3} \, [rad]$$
$$\Delta_p = \theta_p \times H_{eff} = 4.03 \times 10^{-3} \times 0.75 \times 15.3 = 4.60 \times 10^{-2} \, [m]$$
$$\Delta_{total} = \Delta_y + \Delta_p = 8.60 \times 10^{-2} \, [m]$$

Applying the above information, the corresponding displacement ductility is achieved equal to 2.15. Consequently, the median intensity is calculated as follows.

$$\frac{S_a}{S_{ay}} = 1 + (\mu - 1)^{\frac{1}{b}} = 1 + (2.15 - 1)^{\frac{1}{1.23}} = 2.12$$
$$S_a = 0.43 \, g$$

In the process of assessment, the next step is approximation of the annual probability of curvature capacity exceedance. For that purpose, the *b* value is found from Table 1 after estimating period as illustrated in Equation (1); the hazard second order fit coefficients have been found to be $k_0 = 8.54 \times 10^{-4}$, $k_1 = 148.95 \times 10^{-2}$, and $k_2 = 5.78 \times 10^{-2}$, as illustrated in Figure A2. These parameters are implemented in Equations (2)–(5) to compute the annual probability of exceeding the limit state (as presented in Table 8) associated with the given mechanism. The required hand calculation is presented below for illustration.

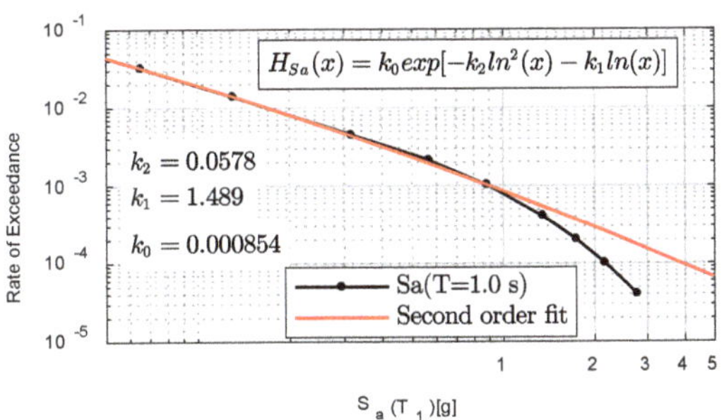

Figure A2. Wellington 1.0 s spectral acceleration hazard curve and fit curve as per recommendation by Vamvatsikos [16].

$$H(S_a) = k_0\left(-k_2 ln^2 S_a - k_1 ln\, S_a\right); S_a = 0.43\ g$$
$$8.54 \times 10^{-4}\left(-5.78 \times 10^{-2} ln^2 0.43 - 148.95 \times 10^{-2} ln\, 0.43\right) = 29 \times 10^{-4}$$
$$p = \frac{1}{1 + 2k_2\left(\beta_{Tot}^2\right)} = \frac{1}{1 + 2 \times 5.78 \times 10^{-2}(0.50^2)} = 97.2 \times 10^{-2}$$
$$P_{LS} = \sqrt{p}\,k_0^{1-p}[H(S_a)]^p \exp\left[\frac{k_1^2}{4k_2}(1-p)\right]$$
$$\sqrt{0.972} \times (8.54)^{1-0.972}\left[29 \times 10^{-4}\right]^{0.972} exp\left[\frac{\left(148.95 \times 10^{-2}\right)^2}{4 \times 5.78 \times 10^{-2}}(1-0.972)\right] = 0.0036$$

Following the above calculation, the annual probability of exceeding the curvature mechanism limit state is found to be 0.0036, which is also reported in Table 7.

References

1. CEN. *Eurocode 8: Design of Structures for Earthquake Resistance-Part 1: General Rules, Seismic Actions and Rules for Buildings*; CEN: Brussels, Belgium, 2005.
2. FEMA. *NEHRP Guidelines for the Seismic Rehabilitation of Buildings and Commentary*; FEMA: Washington, DC, USA, 1997; pp. 273–274.
3. NZSEE. *The Seismic Assessment of Existing Buildings: Technical Guidelines for Engineering Assessments*; Part C-Detailed Seismic Assessment; NZSEE: Wellington, New Zealand, 2017.
4. Sullivan, T.J. Use of limit state loss versus intensity models for simplified estimation of expected annual loss. *J. Earthq. Eng.* **2016**, *20*, 954–974. [CrossRef]
5. Sullivan, T.J.; Welch, D.P.; Calvi, G.M. Simplified seismic performance assessment and implications for seismic design. *Earthq. Eng. Eng. Vib.* **2014**, *13*, 95–122. [CrossRef]
6. Welch, D.P.; Sullivan, T.J.; Calvik, G.M. Developing direct displacement-based procedures for simplified loss assessment in performance-based earthquake engineering. *J. Earthq. Eng.* **2014**, *18*, 290–322. [CrossRef]
7. Vamvatsikos, D.; Aschheim, M.A. Performance-based seismic design via yield frequency spectra. *J. Earthq. Eng. Struct. Dyn.* **2016**, *45*, 1759–1778. [CrossRef]
8. Priestley, M.; Calvi, G.M.; Kowalsky, M. Direct displacement-based seismic design of structures. In Proceedings of the NZSEE conference, Convention Centre, Palmerston North, New Zealand, 30 March 2007.
9. Cornell, C.A.; Jalayer, F.; Hamburger, R.O.; Foutch, D.A. Probabilistic basis for 2000 SAC federal emergency management agency steel moment frame guidelines. *J. Struct. Eng.* **2002**, *128*, 526–533. [CrossRef]
10. Fajfar, P.; Dolšek, M. A practice-oriented estimation of the failure probability of building structures. *J. Earthq. Eng. Struct. Dyn.* **2012**, *41*, 531–547. [CrossRef]
11. Priestley, M.N.; Calvi, G.M.; Kowalsky, M.J. *Displacement-Based Seismic Design of Structures*; IUSS Press: Pavia, Italy, 2007.
12. Orumiyehei, A.; Sullivan, T.J. Quantifying the Likelihood of Exceeding a Limit State via the Displacement-based Assessment Approach. *J. Earthq. Eng.* **2021**, 1–19. [CrossRef]

13. Sullivan, T.J.; Calvi, G.M. *Developments in the Field of Displacement-Based Seismic Assessment, Research Report ROSE 2013/01*; IUSS Press: Pavia, Italy, January 2013; pp. 1–531. ISBN 978-88-6198-090-7.

14. Cardone, D.; Flora, A. Multiple inelastic mechanisms analysis (MIMA): A simplified method for the estimation of the seismic response of RC frame buildings. *J. Eng. Struct.* **2017**, *145*, 368–380. [CrossRef]

15. Pampanin, S.; Bolognini, D.; Pavese, A. Performance-based seismic retrofit strategy for existing reinforced concrete frame systems using fiber-reinforced polymer composites. *J. Compos. Struct.* **2007**, *11*, 211–226. [CrossRef]

16. Vamvatsikos, D. Derivation of new SAC/FEMA performance evaluation solutions with second-order hazard approximation. *J. Earthq. Eng. Struct. Dyn.* **2013**, *42*, 1171–1188. [CrossRef]

17. Bakalis, K.; Vamvatsikos, D. Seismic Fragility Functions via Nonlinear Response History Analysis. *J. Struct. Eng.* **2018**, *144*, 04018181. [CrossRef]

18. Veletsos, A.S.; Newmark, N.M. Effect of inelastic behavior on the response of simple systems to earthquake motions. In Proceedings of the 2nd World Conference Earthquake Engineering, Tokyo, Japan, 11–18 July 1960; pp. 895–912.

19. Vamvatsikos, D.; Cornell, C.A. Incremental dynamic analysis. *Earthq. Eng. Struct. Dyn.* **2002**, *31*, 491–514. [CrossRef]

20. Jalayer, F.; Cornell, C. Alternative non-linear demand estimation methods for probability-based seismic assessments. *J. Earthq. Eng. Struct. Dyn.* **2009**, *38*, 951–972. [CrossRef]

21. Freeman, S. (Ed.) Evaluations of existing buildings for seismic risk-A case study of Puget Sound Naval Shipyard. In Proceedings of the 1st U.S. National Conference on Earthquake Engineering, Bremerton, WA, USA, 18–20 June 1975.

22. Fajfar, P.; Gašperšič, P.; Drobnič, D. Seismic Design Methodologies for the Next Generation of Codes. In *A Simplified Nonlinear Method for Seismic Damage Analysis of Structures*; Routledge: Rotterdam, The Netherland, 1997.

23. Federal Emergency Management Agency—FEMA. *Recommended Seismic Design Criteria for New Steel Moment-Frame Buildings*; Rep. No. FEMA-350; SAC Joint Venture: Washington, DC, USA, 2000.

24. Jalayer, F.; Cornell, C.A. *A Technical Framework for Probability-Based Demand and Capacity Factor (DCFD) Seismic Formats*; Report No. RMS-RMS Program; PEER: Stanford, CA, USA, 2003.

25. Miranda, E. Evaluation of site-dependent inelastic seismic design spectra. *J. Struct. Eng.* **1993**, *119*, 1319–1338. [CrossRef]

26. Jalayer, F. Direct Probabilistic Seismic Anaysis: Implementing Non-Linear Dynamic Assessments. Ph.D. Thesis, Stanford University, Stanford, CA, USA, 2003.

27. Stafford, P.J.; Sullivan, T.J.; Pennucci, D. Empirical correlation between inelastic and elastic spectral displacement demands. *Earthq. Spectra* **2016**, *32*, 1419–1448. [CrossRef]

28. Priestley, M.N.; Calvi, G.M. Towards a capacity-design assessment procedure for reinforced concrete frames. *Earthq. Spectra* **1991**, *7*, 413–437. [CrossRef]

29. Beyer, K.; Dazio, A.; Priestley, M.N. *Seismic Design of Torsionally Eccentric Buildings with U-Shaped RC Walls*; IUSS Press: Pavia, Italy, 2008.

30. Pennucci, D.; Sullivan, T.J.; Calvi, G.M. Inelastic higher-mode response in reinforced concrete wall structures. *Earthq. Spectra* **2015**, *31*, 1493–1514. [CrossRef]

31. Fox, M.J.; Sullivan, T.J.; Beyer, K. Evaluation of seismic assessment procedures for determining deformation demands in RC wall buildings. *J. Earthq. Struct.* **2015**, *9*, 911–936. [CrossRef]

32. SANZ. *Concrete Structures Standard*; NZS 3101; Standards Association: Wellington, NZ, USA, 2006.

33. Sullivan, T.J.; Salawdeh, S.; Pecker, A.; Corigliano, M.; Calvi, G.M. Soil-foundation-structure interaction considerations for performance-based design of RC wall structures on shallow foundations. In *Soil-Foundation-Structure Interaction*; CRC Press: Leiden, The Netherlands, 2010; pp. 193–200.

34. Millen, M.D.; Pampanin, S.; Cubrinovski, M. An integrated performance-based design framework for building-foundation systems. *J. Earthq. Eng. Struct. Dyn.* **2021**, *50*, 718–735. [CrossRef]

35. Sullivan, T.J.; Priestley, M.N.; Calvi, G.M. *A Model Code for the Displacement-Based Seismic Design of Structures DBD12*; IUSS Press: Pavia, Italy, 2012.

36. Lupoi, G.; Franchin, P.; Lupoi, A.; Pinto, P.E. Seismic fragility analysis of structural systems. *J. Eng. Mech.* **2006**, *132*, 385–395. [CrossRef]

37. Carr, A. *RUAUMOKO, Software for Inelastic Dynamic Analysis*; Department of Civil Engineering, University of Canterbury: Christchurch, New Zealand, 2021.

38. Yeow, T.; Orumiyehei, A.; Sullivan, T.J.; MacRae, G.; Clifton, G.; Elwood, K. Seismic performance of steel friction connections considering direct-repair costs. *Bull. Earthq. Eng.* **2018**, *16*, 5963–5993. [CrossRef]

39. Bradley, B.A. A generalized conditional intensity measure approach and holistic ground-motion selection. *J. Earthq. Eng. Struct. Dyn.* **2010**, *39*, 1321–1342. [CrossRef]

40. O'Reilly, G.J.; Sullivan, T.J. Quantification of modelling uncertainty in existing Italian RC frames. *J. Earthq. Eng. Struct. Dyn.* **2017**, *47*, 1054–1074. [CrossRef]

41. Gokkaya, B.U.; Baker, J.W.; Deierlein, G.G. Quantifying the impacts of modeling uncertainties on the seismic drift demands and collapse risk of buildings with implications on seismic design checks. *J. Earthq. Eng. Struct. Dyn.* **2016**, *45*, 1661–1683. [CrossRef]

42. MathWorks, I. MATLAB: The Language of Technical Computing; Desktop Tools and Development Environment. Version 7. MathWorks: 2005. Available online: https://de.mathworks.com/company/aboutus/policies_statements/patents.html (accessed on 6 July 2021).

43. Orumiyehei, A.; Sullivan, T.J. Assessment of the annual probability of failure for systems susceptible to two correlated failure mechanisms. In Proceedings of the 17th World Conference on Earthquake Engineering, Sendai, Japan, 27 September–2 October 2021.
44. Baker, J.W. Efficient analytical fragility function fitting using dynamic structural analysis. *Earthq. Spectra* **2015**, *31*, 579–599. [CrossRef]
45. Pennucci, D.; Sullivan, T.J.; Calvi, G.M. *Performance-Based Seismic Design of Tall RC Wall Buildings*; Iuss Press: Pavia, Italy, 2011.
46. Priestley, M.N.; Seible, F.; Calvi, G.M. *Seismic Design and Retrofit of Bridges*; John Wiley & Sons: Ho boken, NJ, USA, 1996.

Article

A Simplified Approach for the Seismic Loss Assessment of RC Buildings at Urban Scale: The Case Study of Potenza (Italy)

Amedeo Flora *, Donatello Cardone, Marco Vona and Giuseppe Perrone

School of Engineering, University of Basilicata, Viale dell'Ateneo Lucano 10, 85100 Potenza, Italy; donatello.cardone@unibas.it (D.C.); marco.vona@unibas.it (M.V.); giuseppe.perrone@alice.it (G.P.)
* Correspondence: amedeo.flora@unibas.it

Abstract: Comprehensive methodologies based on a fully probabilistic approach (i.e., the performance-based earthquake engineering approach, PBEE), represent a refined and accurate tool for the seismic performance assessment of structures. However, those procedures are suitable for building-specific evaluations, appearing extremely time-consuming if applied at the urban scale. In the proposed contribution, simplified loss assessment procedure will be applied at the urban scale with reference to the residential building stock of the center of Potenza. After the identification of the main reinforced concrete (RC) structural typologies and the definition of specific archetype building numerical models, the direct estimation of expected annual loss (DEAL) methodology will be applied to derive the EAL (i.e., expected annual loss) of RC buildings, deriving information on the effectively seismic quality (or seismic resilience) of the aforementioned built heritage at urban scale. Similarly, the monetary losses associated with downtime are evaluated. Preliminary considerations on the socio-economic effects of seismic scenarios on the territorial scale are also proposed.

Keywords: seismic performance assessment; direct and indirect losses; RC frame buildings; seismic resilience; socio-economic impacts of seismic scenarios

Citation: Flora, A.; Cardone, D.; Vona, M.; Perrone, G. A Simplified Approach for the Seismic Loss Assessment of RC Buildings at Urban Scale: The Case Study of Potenza (Italy). *Buildings* **2021**, *11*, 142. https://doi.org/10.3390/buildings11040142

Academic Editors: Rita Bento and Ana Simões

Received: 26 February 2021
Accepted: 24 March 2021
Published: 1 April 2021

Publisher's Note: MDPI stays neutral with regard to jurisdictional claims in published maps and institutional affiliations.

1. Introduction

Generally speaking, the seismic assessment of a building represents a powerful tool for (i) the evaluation of potential negative effects of significant seismic events occurring on a specific structure and (ii) the definition of the relevant strategies for the seismic retrofitting. The mentioned negative effects belong to different categories such as: structural and non-structural (physical) damage, direct losses associated with such damage (repair costs or reconstruction), indirect (economic) losses connected to building downtime (loss of productivity, business interruption, cost of occupants reallocation), and casualties. On the other hand, the choice of the better retrofit strategy can be properly defined based on rational criteria and building-specific considerations driven by the seismic assessment results [1–3]. In the last decades, several methodologies for the performance based-loss assessment have been proposed [4]. Among those, performance-based earthquake engineering (PBEE) methodology [5] represents the most comprehensive approach providing probabilistic estimates of seismic losses (direct and indirect). The principal output parameter of the mentioned approach is represented by the expected annual loss (EAL) [6], defined as the average economic loss expected to accrue every year in the structure, considering both the direct repair costs related to damage and the social costs associated to downtime (relocation and business interruption). Generally speaking, the lower is the value of EAL, the higher is the seismic quality of examined building. Obviously, the synthetic information expressed by these parameters represent a powerful tool for the stakeholders. In particular, the EAL estimations in both, the as-built and retrofitted configurations can be assumed as input parameters for a cost–benefit analysis, aimed at the rational choice of the optimal retrofit intervention among different options. The accurate evaluation of EAL requests the direct integration, over the site hazard curve, of the monetary losses conditioned on the assumed intensity measure (IM) [7]. As a consequence, single intensity-based loss estimates, performed based on the results of non-linear dynamic analysis, are needed. Specific tools for the practical application of the PEER methodology have been proposed. The performance assessment calculation tool (PACT) represents a valuable mean for the rigorous estimation of EAL. However, it is worth noting that a correct application of this tool requires advanced skills in the non-linear modeling of structures as well as the definition of proper fragility functions and a detailed inventory of damageable components.

The described potentialities of an accurate seismic assessment appears even more attractive in the optic of an urban scale evaluation. Indeed, in this case, the derived synthetic global parameters (namely EAL) and the results of the eventual cost–benefit analyses could be used by the decision makers (government, public institutions, and organizations such as civil protection, etc.) to properly define the direction of their political choices. To better support decision makers, several studies have been carried out with the use of multi-criteria decision methods (MCDMs). In recent studies, MCDMs have been widely used particularly to define a prioritization list and for the selection of the optimal intervention strategies [8]. The mentioned results, together with further assessments on the potential social and economic large scale impacts (such as in the loss of the quality of life of the population, performance reduction of the critical infrastructures, etc.), are crucial elements to perform a resilience analysis as well as a pre- and post-event management at urban scale [9].

However, based on previous considerations, the inherent complexity of the PEER–PBEE methodology became unaffordable in case of an urban scale evaluation. In other words, the application of an accurate building-specific approach to all the structures composing the building stock appears computationally expensive, in particular for practitioners, and excessively time-consuming. Even classifying the structures located on a certain area in a limited number of typologies, the modeling effort remains particularly onerous, as well as the amount of input data for a proper application of PACT process.

Recently, many authors proposed simplified approaches relating the economic losses to the structural response (engineering demand parameters, EDP) derived from analysis methods within the reach of the most practicing engineers. In particular, the main simplification lies in relating a given performance level (PL) to an expected economic loss pre-defined based on the actual structural typology [10] or calculated based on specific story-based loss function [11]. The first approach is currently used within the Italian Seismic Risk Classification [12] to evaluate the "seismic quality" of existing buildings, providing a specific score system articulated in eight seismic risk classes. However, as observed by Perrone et al. (2019) [13], this approach is generally affected by a significant loss of accuracy in many cases, in particular for buildings characterized by drift/acceleration profiles or structural/non structural elements distributions sensibly different in the two main directions or along the building height. In contrast, the accurateness of the results derived using procedures based on the definition of story-based loss functions appears satisfactory [13]. In this context, Cardone et al. (2017) [14] firstly proposed a new approach, later developed by Perrone et al. 2019 [13], referred to as direct estimation of expected annual losses (DEAL). Under specific assumptions, the EAL of a single building can be evaluated using a closed-form equation, significantly reducing the complexity of a large scale seismic assessment.

All that considered, in the present paper, the DEAL approach has been applied at the urban scale for the seismic loss assessment of the city center of Potenza (southern Italy). A series of operative choices have been implemented to operatively perform the seismic assessment. First of all, the main structural and typological peculiarities of the residential RC buildings have been collected to define a general typologies inventory. In the second step, a series of case-study buildings (archetypes), each one associated with a single typology, have been defined and modeled using the Opensees Framework [15]. Successively, non-linear static analyses have been carried out on the mentioned non-linear models to derive the structural response, hence the main EDPs, representing the input data for the application of the DEAL method, aimed at the evaluation of the EAL associated with direct losses. In the last part of the paper, an estimation of the EAL associated with indirect losses related to downtime is proposed according to the methodology presented in Cardone et al. (2019) [16]. Finally, based on the results in terms of monetary losses, preliminary considerations on the economic and social impacts of probable seismic scenarios are made. The main novelty of the present study is represented by the application of the DEAL methodology within an integrated approach, affordable for skilled practitioners and also useful to derive territorial scale considerations at different levels (stakeholders, local government, civil protection, etc.).

2. Overview of the Direct Estimation of Expected Annual Loss

The displacement-based (DB) simplified seismic loss assessment approach proposed by Cardone et al. (2020) [4] for RC structures is characterized by the following key points: (i) estimation of EDPs (i.e., floor acceleration and/or inter-story drifts profiles) through simple analysis methods, (ii) application of proper story-based loss functions for the estimation of expected losses at different performance levels, and (iii) direct calculation of EAL through a closed-form equation.

Two main assumptions have been considered in this approach: (i) linear increase of the expected monetary losses (associated with the seismic event) with the IM and (ii) linear approximation of the seismic hazard curve in log-space [17] (see Figure 1a).

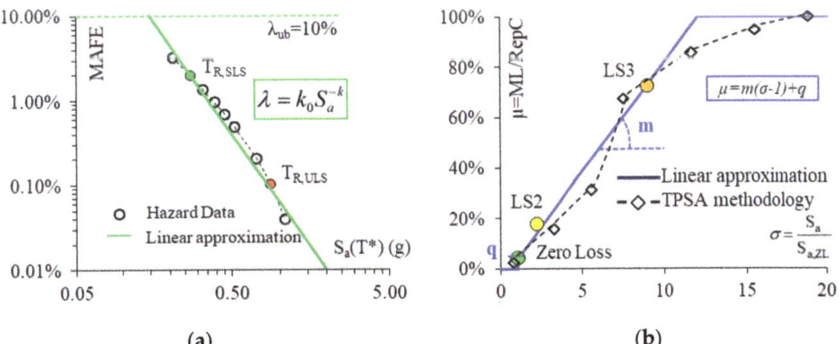

Figure 1. Assumptions of the proposed simplified procedure [4]: (**a**) hazard curve linear approximation and (**b**) expected loss vs. IM relationship.

More in detail, the hazard curve describes the mean annual frequency of exceedance (MAFE) corresponding to a certain level of ground motion as a function of the selected intensity measure, namely the spectral acceleration at the fundamental period of vibration, $S_a(T^*)$. It is worth noting that T^* is assumed equal to the mean period of vibration in the two orthogonal directions of the structure. The following relationship is used to describe the linear approximation of the hazard curve:

$$\lambda = k_0 S_a(T^*)^{-k}, \tag{1}$$

where:

$$k = [Log(\lambda_{ULS}/\lambda_{SLS})]/[Log(S_{a,ULS}/S_{a,SLS})], \tag{2}$$

$$k_0 = \lambda_{SLS} \times S^k_{a,SLS}, \tag{3}$$

with λ_{SLS} and λ_{ULS} representing the reciprocal of the return period, T_R, of earthquake at the serviceability limit state (SLS) and ultimate limit state (ULS) assumed by the current seismic code [18] and $S_{a,SLS}$ and $S_{a,ULS}$ representing the associated spectral accelerations at T^*. In order to align the DEAL approach with that provided within the Italian seismic classification guidelines [12], a maximum value of the MAFE, namely $\lambda_{ub} = 10\%$ (return period equal to 10 years) has been assumed in the procedure.

As showed in [13], expected monetary losses increase almost linearly compared to the IM. In Figure 1b the loss curve of a reference RC frame building (dash-dotted line) is normalized on both axes [19]. In particular, expected monetary losses are normalized with respect to the replacement cost (RepC) of the building, while the spectral acceleration to the zero-loss seismic intensity, $S_{a,ZL}$ (IM), corresponds to a minor damage that is not going to be repaired [14]). The loss curve of the mentioned reference RC frame building has been obtained by applying the accurate time-based performance seismic assessment (TPSA) methodology proposed within FEMA P-58 [20].

All that considered, a generic story-based loss curve can be approximated using the following Equation:

$$\mu = m(s - 1) + q \leq 1, \tag{4}$$

In Equation (4), the parameter q is the initial-loss threshold corresponding to the cosmetic damage loss associated with $S_{a,ZL}$. The slope m can be obtained using a best fit regression linear approach considering other two limit states besides the zero loss point: the operational (OP) and damage control (DC) limit states. The values of $S_a(T^*)$ associated with the aforementioned limit states (ZL, OP, and DC) can be evaluated using simplified methodologies as the displacement base assessment (DBA), the MIMA approach [21], etc. On the other hand, the corresponding monetary losses can be obtained using proper story-based loss function, such as those proposed by Cardone et al. (2020) [4] for residential RC buildings. More details about the described approach can be found in [13]. Once the parameters m and q are determined, the following closed form equation can be applied to derive the direct EAL [10]:

$$EAL = \lambda_{min} q_{min} + (k_0 / S^k_{a,ZL}) \times [(m/(1 - k) \times (s^{1-k}_{TL} + s^{1-k}_{min})], \tag{5}$$

in which:

$$q_{min} = m(s_{min} - 1) + q, \tag{6}$$

$$s_{TL} = [(1-q)/m] + 1, \tag{7}$$

$$s_{min} = S_{a,min}/S_{a,ZL} = \max(1; S_{a,ub}/S_{a,ZL}) = \max(1; (\lambda_{ub}/k_0)^{-1/k}/S_{a,ZL}), \tag{8}$$

The parameter σ_{TL} is related to the IM value at which the normalized expected losses are equal to the 100% of RepC, λ_{ub} represents the MAFE maximum limit, σ_{min} is associated with the maximum (normalized) IM between the ZL and the spectral acceleration associated to the MAFE maximum limit ($S_{a,ub}$), and q_{min} is the corresponding monetary loss.

For what concerns the indirect component of EAL, a simplified approach has been proposed in [16]. In particular, this method takes into account the indirect losses directly associated to the interval of time between the seismic event and the end of repair activities, namely the so-called "downtime". Two main components are considered: rational and irrational. The first one represents the time needed to complete the building repair activities. Reference to a "Slow-Track" approach has been made to derive the rational component of downtime. In other words, the repair activities are performed at each floor progressively and independently from one story to another [16]. The second includes several preliminary activities: bureaucratic issues, damage assessment, financial planning, occupants relocation, technical design, etc.

Based on the mentioned approach, indirect losses vs. normalized spectral acceleration $(S_a(T^*)/S_a(T^*)_{ZL}$ relationships can be derived for a specific building. It is worth noting that a series of assumptions have been made regarding the irrational component of downtime and, in particular, regarding the relocation of buildings occupants. More in detail, in the immediate aftermath of the seismic event, building occupants are generally hosted in hotels at sub-sized rates. In a second phase, these occupants usually move to "long term" accommodations (e.g., private apartments or emergency structures). All that considered, the following assumptions have been made: (i) occupants are accommodated in hotels considering an average cost of €40/room/day for no more than 90 days; (ii) average lease rate (C_{rental}) equal to €3.5/m^2/month [22]; (iii) average occupancy ratio equal to 1.4 persons/100 m^2 [22]. However, many factors could modify the mentioned assumptions, thus enhancing the level of uncertainty of the procedure. For example, based on the effective statistics provided by the Italian Government, the number of building occupants that effectively move to a hotel accommodation is generally lower than the total number of occupants (N_{occ}). Indeed, a part of them owns a secondary home (namely, holiday house) or is hosted by relatives. Similarly, several factors could influence the relocation time, leading to a period range variable from few days to several weeks. In this optic, two main scenarios have been defined (i.e., lower bound and upper bound conditions). In the lower bound condition, only 2/3 of the occupants are relocated. In this case, 45 days have been assumed as a relocation time. In the upper bound scenario, the totality of occupants is relocated, assuming a 90-day relocation time.

3. Building Typologies Identification for the City Center of Potenza

Potenza is the main city of the Basilicata region (southern Italy), counting about 65,000 inhabitants. The evolution of the current built heritage of Potenza is strongly related to the historical events occurred during the last two centuries.

During the XIV century, the city was hit by different significant seismic events (intensity higher than VIII MCS). Two close strong earthquakes, occurring in 1826 and 1857, produced several victims and severe damage in the entire town [23]. Important demolition and reconstruction activities were undertaken in the historical city center in the aftermath of these events [23]. In the twentieth century, a massive migration of the population from the historical city center to modern neighborhoods (in the upper west side of the city) was registered due to the demographic increment registered during the 1930s and 1940s. As a consequence, in the early 1950s, the local social housing organization funded and pursued an important public housing plan creating new residential neighborhoods. After the strong earthquake occurred on 23 November 1980 (Irpinia and Basilicata earthquake), a lot of existing buildings were retrofitted [24] using public financial resources allocated within the law 219/81.

In this paper, the seismic assessment is referred to two main areas: the old town center (located in the hilltop town, labeled as C1 in Figure 2) and the residential public housing compartment (labeled as C2). These two compartments feature homogeneous urbanistic, constructive, and historical peculiarities, including a large part of the built heritage of Potenza developed during 1850–1950 and 1945–1990, respectively. In this study, only private RC residential buildings have been considered. As a matter of fact, public buildings have been neglected due to their limited number and their inherent peculiarities deserving a specific study. As mentioned in Section 1, the building-specific approach appears extremely time consuming and computationally unfeasible for regions or areas populated by hundreds of buildings. As a consequence, the definition of a building typologies inventory represents a pragmatic choice for a simplified seismic assessment at urban-scale. Grouping buildings in limited number of typologies is an affordable operation, especially when the urban area can be divided in many sub-areas (namely compartments) characterized by similar urbanistic, historical, and constructive peculiarities.

Several criteria can be assumed for the structural typologies identification and distribution on a given area. Obviously, the consistency and completeness of the identification approach is one of the main elements affecting the accuracy of the seismic assessment. Elementary building typologies inventories are fundamentally based on the structural system construction material [25]. More advanced approaches include additional information as construction period, non-structural elements characteristics (materials and sizes), primary load bearing structure, number of stories, etc. [26]. Recently, very accurate building classification methodologies have been proposed and also incorporated in several seismic codes [27]. However, the presence and spread of a construction typology on a certain territory is generally related to many different factors, strictly connected to the inherent peculiarities of a territory: socio-economic conditions, geological and topographical conditions, traditional building technologies, etc. In this optic, customized approaches represent a proper solution for a reliable buildings classification.

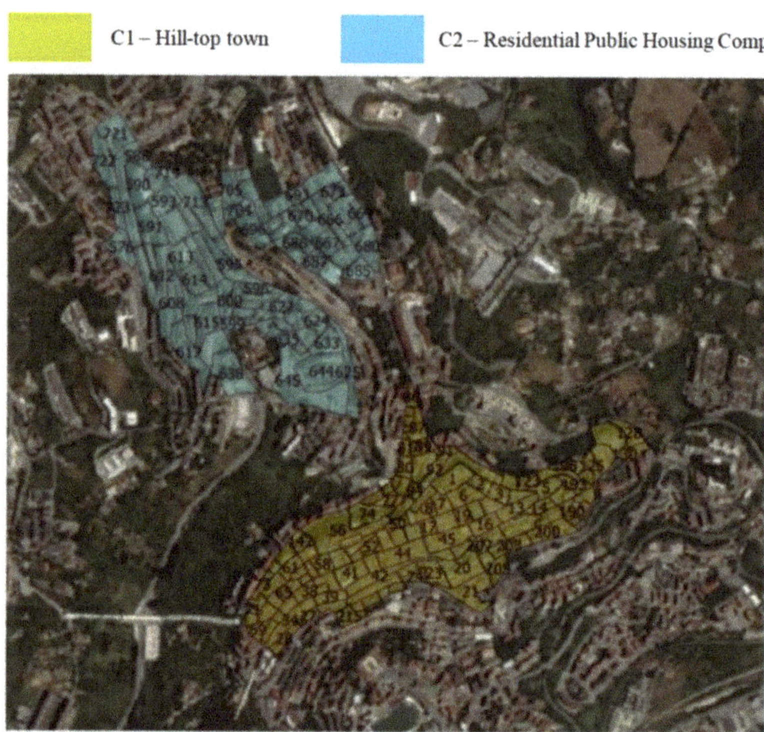

Figure 2. Compartments overview and census tracts: (yellow) old town center and (blue) residential compartments.

A typical approach for a customized classification of building typologies at territorial-scale is based on the assumption of the census database as primary source of information. Subsequently, these data (for each census tract) are integrated with other specific sources of information (in situ and virtual inspections, documental analysis, etc.) [28]. The heterogeneous data supplied by the Italian Institute of Statistics (ISTAT, [22]) appears sufficiently populated (statistics updated every 10 years) and homogeneously distributed. In Table 1, the most relevant data referring to the examined compartments derived by the ISTAT Buildings Database are summarized [22].

Table 1. Relevant data for the examined compartments.

Compartment	n° of Buildings	Masonry (Nr.(%))	RC (Nr.(%))	Pre-81 (%)	Post-81 (%)	Ns $\leq$ 3 (%)	Ns > 3 (%)
C1	429	327 (76%)	102 (24%)	90	10	49	51
C2	128	88 (69%)	40 (31%)	90	10	22	78

Based on the synthetic data reported above, two main elements emerged as significant classification variables: the number of stories (n_s) and the age of construction (a_c). Considering the aggregated form of the census data, no details are provided to directly derive more specific information as the number of RC buildings showing two, three, or more than three stories or, similarly, how many RC structures have been realized in a certain period. For what concerns the number of stories, in order to overcome this limitation, innovative approaches, namely G.I.S. technology and high resolution (HR) optical satellite imagery, have been adopted to collect more detailed geo-referenced data on the effective height of the RC buildings in the examined territory [28], thus obtaining an adequate disaggregation of primary source data (census). With reference to n_s, three macro-classes of buildings have been identified: "Lr" (from one to three stories), "Mr" (from four to six stories), and "Hr" (from seven to 10 stories). The criterion adopted for the definition of the mentioned number of stories ranges (1–3, 4–6, and 7–10) is in line with the observation presented in Masi et al. [29]. As a matter of fact, all other variables being equal, RC buildings featuring one to three stories exhibit a similar seismic behavior [29]. The same considerations can be made for medium-rise RC buildings (four to six stories) and for high-rise buildings (seven to 10 stories).

On the other hand, two main classes have been defined with reference to the age of construction: pre- and post-1981. The identification of these classes is not merely temporal, but is associated to the reference design code. Indeed, the introduction of new seismic regulations after the Irpinia Earthquake led to enhanced anti-seismic peculiarities for buildings designed after 1981. An extensive documental investigation has been performed for the identification of the main structural peculiarities of buildings realized pre- and post-1981. A significant database furnished by the Regional Public Social Housing Organization (ATER), directly involved in the post-seismic residential reconstruction from 1981 to 1990, has been analyzed. Moreover, building inspections have been conducted gathering fundamental data on dimensional and structural peculiarities of structures sited in the examined areas.

The main structural characteristics of pre- and post-1981 buildings are resumed in what follows. For pre-1981 buildings, the typical lateral resisting system is characterized by perimeter and internal resisting frames along a single direction. Moreover, internal shallow beams and external deep beams were generally adopted. Resisting frames along both principal directions and a large use of shallow beams have been observed for post-1981 buildings. For what concerns non-structural elements, heavy masonry infills, constituted by a double layer (external 13 cm solid bricks and internal 10 cm hollow clay bricks with a 10 cm inter-space) have been observed for most of pre-1981 structures, in particular for those realized before 1961. In a few cases (in particular for buildings realized during the period 1960–1980) light masonry infills (hollow bricks positioned in two layers of 10 + 10 cm with a 10 cm inter-space) have been detected. The same configuration has been observed for post-1981 buildings.

The combination of the described peculiarities for each building class lead to a different seismic vulnerability. In this optic, a preliminary classification, in terms of seismic vulnerability, has been defined. Preliminary macro-typologies have been considered, featuring (i) low (L), (ii) medium (M), and (iii) high (H) vulnerability (see Table 2). Subsequently, the preliminary classification has been refined including specific attributes that directly affect the structural behavior. In particular, vertical irregularities and staircase typology have been taken into account. Irregularities in elevation have been observed for pre-1981 (in particular for those realized in the 1970s) and post-1981 buildings. Pilotis stories (namely open ground story) are frequent in particular in the C2 compartment. Moreover, two staircase typologies have been identified: knee beams with cantilever steps (labeled as *k* in what follows) represent the typical solution for buildings realized before 1981 and for a large part of post-1981 buildings; Waist-slab staircases (labeled as *s*) have been detected in a number cases for post-1981 buildings (in particular for those realized after 1990). The structural configurations (in plan and elevation) and the effective structural element dimensions observed during the described investigations have been properly considered in the numerical modeling (see Section 3). Negligible differences in terms of horizontal floor type, roof typology, and state of preservation have been observed.

Table 2. Preliminary macro-typologies classification.

ID	Description
L (Low)	Bi-directional lateral resisting system Light masonry infills Seismic Resistant Design
M (Medium)	Mono-directional resisting system Light masonry infills Gravity load design
H (High)	Mono- directional resisting system Heavy masonry infills Gravity load design

The final typologies inventory, resulting from the classification approach discussed above, is reported in Table 3. The theoretical total number of classes is equal to 36. However, the number of buildings effectively included in some of those classes is extremely limited. Consequently, to reduce the computational amount of the procedure and considering the low incidence of a limited number of elements on the global seismic assessment of the area, the typologies including a number of buildings lower than three have been neglected in what follows. Finally, eight typologies of the defined inventory have been effectively taken into account counting 131 buildings, distributed as reported in Table 4, thus covering the 92% of the RC residential buildings population. In other words two "virtual" compartments consisting only of residential RC buildings (afferent to the previously defined typologies) have been generated.

Table 3. Building typologies inventory.

Macro Typology			Nr. of Stories			Staircase Typolgy		Vertical Irregularities		ID
L	M	H	Lr	Mr	Hr	k	s	PF	IF	
X			X			X		X		L, Lr, k, PF
X			X			X			X	L, Lr, k, IF
X			X				X	X		L, Lr, s, PF
X			X				X		X	L, Lr, s, IF
X				X		X		X		L, Mr, k, PF
X				X		X			X	L, Mr, k, IF
X				X			X	X		L, Mr, s, PF
X				X			X		X	L, Mr, s, IF
X					X	X		X		L, Hr, k, PF
X					X	X			X	L, Hr, k, IF
X					X		X	X		L, Hr, s, PF
X					X		X		X	L, Hr, s, IF
	X		X			X		X		M, Lr, k, PF
	X		X			X			X	M, Lr, k, IF
	X		X				X	X		M, Lr, s, PF
	X		X				X		X	M, Lr, s, IF
	X			X		X		X		M, Mr, k, PF
	X			X		X			X	M, Mr, k, IF
	X			X			X	X		M, Mr, s, PF
	X			X			X		X	M, Mr, s, IF
	X				X	X		X		M, Hr, k, PF
	X				X	X			X	M, Hr, k, IF
	X				X		X	X		M, Hr, s, PF
	X				X		X		X	M, Hr, s, IF

Table 3. *Cont.*

Macro Typology			Nr. of Stories			Staircase Typolgy		Vertical Irregularities		ID
L	M	H	Lr	Mr	Hr	k	s	PF	IF	
		X	X			X		X		H, Lr, k, PF
		X	X			X			X	H, Lr, k, IF
		X	X				X	X		H, Lr, s, PF
		X	X				X		X	H, Lr, s, IF
		X		X		X		X		H, Mr, k, PF
		X		X		X			X	H, Mr, k, IF
		X		X			X	X		H, Mr, s, PF
		X		X			X		X	H, Mr, s, IF
		X			X	X		X		H, Hr, k, PF
		X			X	X			X	H, Hr, k, IF
		X			X		X	X		H, Hr, s, PF
		X			X		X		X	H, Hr, s, IF

Table 4. Prevalent building typologies distribution on the examined area.

Building Typology	Number of Elements	RepC (€)
L, Hr, o, IF	7	2,858,000
L, Mr, s, IF	5	2,135,000
M, Hr, k, IF	25	2,858,000
M, Mr, k, IF	22	2,135,000
M, Lr, k, IF	6	1,077,640
H, Hr, k, IF	14	2,858,000
H, Mr, k, PF	12	2,135,000
H, Mr, k, IF	40	2,135,000
Total	131	-

4. Archetype Buildings

4.1. Overview

Once the prevalent building typologies have been detected for the two compartments, a series of archetype buildings, one for each typology, is defined. Such archetypes represent, on average, the geometrical, material, and structural peculiarities of the corresponding typological class. Considering that few differences have been registered in terms of building layout, the same plan configuration has been assumed for all the archetypes (see Figure 3). The floor area is equal to 418 m^2, hosting two apartments per floor. The inter-story height is equal to 3.3 m (3.8 m at first story) for H- and M-archetypes, while is equal to 3.0 m for the L-archetypes. Table 5 summarizes the main geometrical characteristics and steel reinforcements details derived from the documental analyses and in situ inspections described in Section 2. The replacement cost (RepC) of the examined archetypes is shown in the last column of Table 4. It has been estimated considering the average cost of construction per square meter of similar new buildings (€730/m^2 according with CIAMI 2014 [30]). Moreover, a further cost for demolition and disposal of materials equal to €135/m^2 [30] has been taken into account, leading to a total replacement cost equal to approximately €865/m^2.

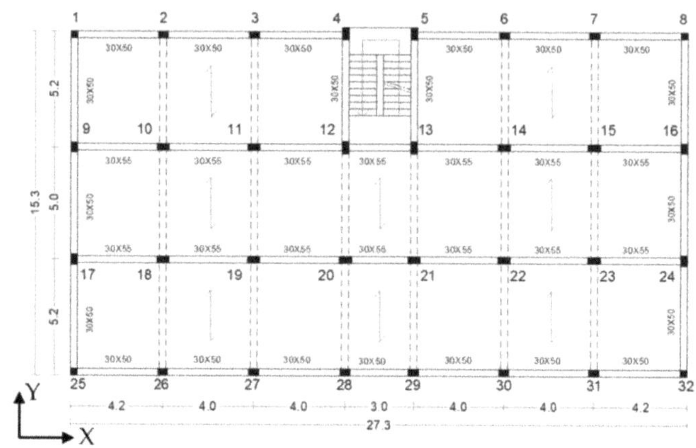

Figure 3. Building layout: structural plan view.

Table 5. Main geometrical characteristics and steel reinforcement details of the selected archetype buildings.

Archetype	Nr. of Frames	Column Section (mm)	Beam Section (mm)	Long. Reinforcements Ratio	Transv. Reinforcements Ratio	Reinforcement Type	Masonry Infills
L, Hr, s, IF	E (X): 2 I (X): 2 E (Y): 2 I (Y): 6	E: 300 × 300–300 × 550 I: 350 × 300–650 × 300 SC: 300 × 650 Cr: 300 × 300	E (X): 300 × 500 I (X): 300 × 550 E (Y): 300 × 500 I (Y): 300 × 400 KB: 300 × 500	B: 0.54–1.07% C: 0.59–1.19%	B: Ø8/150 mm C: Ø8/150 mm SC: Ø8/150 mm	deformed (FeB44k)	100 + 100 mm
L, Mr, s, IF	E (X): 2 I (X): 2 E (Y): 2 I (Y): 6	E: 300 × 300–300 × 450 I: 350 × 300–550 × 300 SC: 300 × 550 Cr: 300 × 300	E (): 300 × 500 I (X): 300 × 550 E (Y): 300 × 500 I (Y): 300 × 400 KB: 300 × 500	B: 0.54–0.94% C: 0.59–1.10%	B: Ø8/150 mm C: Ø8/150 mm SC: Ø8/150 mm	deformed (FeB44k)	100 + 100 mm
M, Hr, k, IF	E (X): 2 I (X): 2 E (Y): 2 I (Y): 0	E: 300 ×3 00–300 × 550 I: 350 × 300–650 × 300 SC: 300 × 650 Cr: 300 × 300	E (X): 300 × 500 I (X): 300 × 550 E (Y):300 × 500 KB: 300 × 500	B: 0.31–0.72% C: 0.58–0.75%	B: Ø6/200 mm C: Ø6/200 mm SC: Ø6/200 mm	smooth (Aq50)	100 + 100 mm
M, Mr, k, IF	E (X): 2 I (dir X): 2 E (Y): 2 I (Y): 0	E: 300 × 300–300 × 450 I: 350 × 300–550 × 300 SC: 300 × 550 Cr: 300 × 300	E (X):300 × 500 I (X):300 × 550 E (Y):300 × 500 KB: 300 × 500	B: 0.31–0.62% C: 0.58–0.70%	B: Ø6/200 mm C: Ø6/200 mm SC: Ø6/200 mm	smooth (Aq50)	100 + 100 mm
M, Lr, k, IF	E (X): 2 I (X): 2 E (Y): 2 I (Y): 0	E: 300 × 300–300 × 350 I: 350 × 300–450 × 300 SC: 300 × 550 Cr: 300 × 300	E (X): 300 × 500 I (X): 300 × 550 E (Y): 300 × 500 KB: 300 × 500	B: 0.31–0.41% C: 0.58–0.68%	B: Ø6/200 mm C: Ø6/200 mm SC: Ø6/200 mm	smooth (Aq50)	100 + 100 mm
H, Hr, k, IF	E (X): 2 I (X): 2 E (Y): 2 I (Y): 0	E: 300 × 300–300 × 550 I: 350 × 300–650 × 300 SC: 300 × 550 Cr: 300 ×3 00	E (X): 300 × 500 I (X): 300 × 550 E (Y): 300 × 500 KB: 300 × 500	B: 0.31–0.72% C: 0.58–0.74%	B: Ø6/250 mm C: Ø6/250 mm SC: Ø6/250 mm	smooth (Aq42)	130 + 100 mm
H, Mr, k, PF	E (X): 2 I (X): 2 E (Y): 2 I (Y): 0	E: 300 × 300–300 × 450 I: 350 × 300–550 × 300 SC: 300 × 500 Cr: 300 × 300	E (X): 300 × 500 I (X): 300 × 550 E (Y): 300 × 500 KB: 300 × 500	B: 0.30–0.62% C: 0.58–0.68%	B: Ø6/250 mm C: Ø6/250 mm SC: Ø6/250 mm	smooth (Aq42)	130 + 100 mm
H, Mr, k, IF	E (X): 2 I (X): 2 E (Y): 2 I (Y): 0	E: 300 × 300 I: 300 × 300 SC: 300 × 300 Cr: 300 × 300	E (X): 250 × 450 I (X): 250 × 450 E (Y): 250 × 450 KB: 250 × 550	B: 0.30–0.60% C: 0.50–0.66%	B: Ø6/250 mm C: Ø6/250 mm SC: Ø6/250 mm	smooth (Aq42)	130 + 100 mm

E: External, I: Internal, Cr: Corner, SC: Staircase; KB: Knee beams, B: Beams; C: Columns.

For what concerns the mechanical properties of concrete, a proper statistical analysis has been carried out. The latter is based on a specific database obtained by assembling the results of compression tests performed by the practitioners performing the seismic assessment on residential RC buildings realized between 1950 and 1990 and located in the Potenza district. The ultimate strength represents the main output data of the compression tests on core specimens. Unreliable or unclear data have been discarded. As observed by Vona (2014) [31], the value of the compressive strength evaluated during the tests (f_{core}) could be sensibly different from the effective in-situ value (f_c), due to different factors, namely, presence of steel bars in the examined concrete portion, ratio between height (h) and diameter (D) of the examined specimen, sample damage, etc. All that considered, in first approximation, the sample's core strength values have been converted to in-situ values using the following relationship:

$$f_c = (M_{,h/D} \times M_{,dia} \times M_{,a} \times M_{,d}) \times f_{core}, \qquad (9)$$

where:

- $M_{h/D}$ represents the modification factor related to the h/D ratio, equal to $M_{,h/D} = 2/(1.5 + D/h)$;
- $M_{,dia}$ represents the diameter modification factor (1.06 for D = 50 mm, 1.00 for D = 100 mm and 0.98 for D = 150 mm);
- $M_{,a}$ is the modification factor related to the presence of steel bars (ranging from 1.03 to 1.13 as a function of the bar diameter) [32];
- $M_{,d}$ is the modification factor accounting for damage occurring during the extraction activities, equal to 1.06 [33].

Subsequently, the aggregated strength values of the database have been disaggregated, grouping the data in four classes, associated with as many construction periods (pre-1961, 1961–1971, 1972–1981, and post-1981), interspersed with important modifications in the building regulations in force in Italy. The changes in the regulations led inevitably to significant modifications also in the materials quality. In Table 6, the results of the statistical evaluation in terms of concrete strength and in the different construction periods are summarized. For what concerns the steel rebars, the information gathered during the documental analysis has been integrated with the database provided by the STIL software [34] to derive the mechanical properties of steel rebars in the mentioned construction periods. The STIL database is based on the results of 19,140 tensile tests performed on steel rebars used in Italy during the period 1950–2000. Selecting the desired period range and providing, as input, the steel type (smooth or deformed) and, eventually, the specific material category (as a function of the production class and/or regulation's classification in force at the reference period), the software returns the corresponding mechanical properties of the material (yield strength, elongation at fracture, hardening ratio). The mean values and the standard deviations of the yield strength (f_y), obtained for the steel rebars in the mentioned construction periods, are reported in Table 7.

Table 6. Mean value and standard deviation of concrete strength (f_c) in different construction periods.

Statistical Values	Construction Period			
	Pre 1961	1961–71	1972–81	Post 1981
Number of specimens	28	132	264	360
Mean value (N/mm^2)	16.32	19.12	22.21	24.73
Standard Deviation	4.34	10.68	11.32	10.59
C.V.	0.25	0.12	0.08	0.07

Table 7. Yield strength (f_y) of steel rebars in different construction periods.

Statistical Parameter	Construction Period			
	Pre 1961	1961–71	1972–81	Post 1981
Mean Value (N/mm^2)	321.2	369.8	433.1	490.3
Standard Deviation	26.8	33.6	32.5	68.6

The disaggregation of census data obtained using secondary sources of information described in Section 2 highlighted that all the buildings afferent to the H macro-typology were realized before 1961. As a consequence, the mechanical properties associated with the latter period have been adopted for the numerical models of the archetype buildings representing this macro-typology. Similarly, the mechanical properties determined for post 1981 buildings have been adopted for those referred to the L macro-typology. Finally, for what concerns the M macro-typology, based on the documental analysis, most of the buildings included in this typology (about 80%) have been realized between 1961 and 1971. Only the 20% was built just before or after the latter decade. As a consequence, for the sake of simplicity, the mean mechanical properties of steel and concrete corresponding to the aforementioned construction period (1962–1971) have been adopted for all the numerical models describing the behavior of M-typology.

For what concerns the seismic hazard, all the archetypes are assumed to be located on a medium-soft soil classified as soil type (C), according with the current Italian Seismic Code [18]. For each archetype, the site hazard curve, defined based on the data provided by the INGV (Italian Institute of Geophysics and Volcanology), is expressed in terms of mean annual frequency of exceedance ($MAFE_i$) as a function of the considered IM, namely, the spectral acceleration ($S_{a,i}(T^*)$), corresponding to the average fundamental period of the structure T^*.

4.2. Numerical Modeling and Analysis Results

Lumped plasticity models have been implemented in the OpenSees framework [15] to describe the non-linear behavior of the archetype buildings defined in the previous section. The structural elements of the resisting frames have been modeled using the Beam With Hinges (BWH) element, already included in OpenSees. This is a force-based element composed by a central linear elastic region and two discrete plastic hinges on both ends. Such plastic hinges feature a non-linear cyclic behavior according to the modified Ibarra-Medina-Krawinkler deterioration model [35]. Moment–curvature analysis of the critical cross sections of the structural elements (beams and columns) have been performed to derive the inherent skeleton curves, also considering the effect of axial load interaction. Reference to Haselton et al. (2015) [36] has been made to define the degradation parameters for strength and post-capping strength deterioration. Bar slipping has been also taken into account for archetype buildings with smooth rebars (i.e., macro classes H and M), by adopting a proper constitutive law for steel, according with Braga et al. (2012) [37]. It is worth noting that, in this case, the contribution of compression longitudinal rebars has been neglected [38]. Moreover, for RC members featuring smooth rebars, a proper value of the plastic hinge length (H/4 and H/3 for base columns and beams, respectively) has been adopted considering that, although the width of flexural cracks strongly increases due to bond slip effects, these cracks do not spread along the element span during repeated cyclic deformations.

The influence of shear failures on the structural response during the analyses has been explicitly taken into account. All the plastic hinges have been pre-qualified as (i) ductile, in which the shear failure is avoided and the moment-rotational backbone model is not modified, or (ii) shear critical, in which the backbone is reduced after the shear failure, following a softening branch up to zero based on the empirical proposal by Aslani and Miranda (2005) [39].

The model adopted for the joint panel zone of exterior unreinforced joints is the so-called scissors model proposed by Alath and Kunnath (1995) [40]. This model appears very simple from a computational point of view, but also sufficiently accurate in describing the experimental beam–column joint behavior of non-ductile RC frames [41]. In particular, the nonlinear behavior of beam–column joints is modeled using rigid offsets connecting the ends of the beams and columns with two nodes (A and B, respectively) overlapped and located in the center of the panel. Nodes A and B are connected through a rotational spring featuring a single degree of freedom (relative rotation) constitutive model, namely Pinching4 uniaxial material, available in the Opensees framework. The latter model is characterized by a quadri-linear moment vs. rotation relationship, directly related to the joint shear stress (τj)–shear strain (γj) behavior (see Figure 4). More details can be found in Ricci et al. (2019) [42]. It is worth noting that a negligible influence of interior beam–column joints was highlighted by preliminary analyses. As a consequence, these nodes are not considered in the final models.

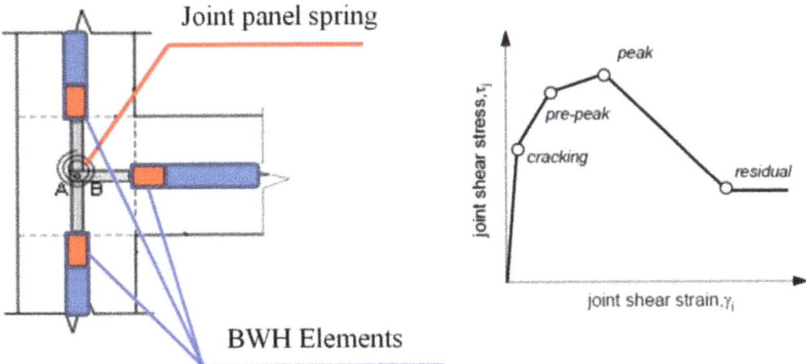

Figure 4. Modeling approach for exterior joint panels.

An equivalent diagonal strut (only in compression) has been adopted to model the masonry infill panels of the examined archetypes. Reference to Sassun et al. (2016) [43] has been made to define the constitutive law of the diagonal struts. In Table 8, the mechanical properties of the constitutive materials composing the (heavy and light) masonry infills are summarized. Specific reduction factors [44] have been considered to opportunely account for the effect of openings, by limiting the strength and lateral stiffness of the panels.

Table 8. Mechanical characteristics of the constitutive materials composing the masonry infill panels.

Brick Type	Mortar	Masonry Panel		
		$_{m0}$ (MPa)	$_{m0}$ (MPa)	E_m (MPa)
Solid bricks thickness 130 mm	Cement + sand	12.00	0.84	6000
Hollow clay bricks thickness 100 mm	Cement + sand	1.20	0.20	1050

The capacity spectrum method (CSM) [45] implementing the N2 approach [46] has been applied to derive the EDPs associated to the reference limit states (ZL, OP, and DC) and the corresponding spectral accelerations at the fundamental period T*. In this optic, non-linear static analyses (push-over) have been performed in the two principal (X- and Y-) directions of each archetype building, considering a linear force distribution (Figures 5–7). It is worth noting that the capacity curves in terms of base shear vs. top displacement have been cut off at a peak strength reduction of about 50% on the negative slope. The failure mode of the H, Mr, k, IF and H, Mr, K, PF archetype buildings is a typical weak-story collapse mechanism located at first story in both cases. However, the differences in terms of columns' dimensions and reinforcement ratios (hence the different response of the involved plastic hinges at first story) produce a different structural behaviour (i.e., different maximum base shear and ultimate displacement) of the two case studies. A mixed-sway mechanism is observed for the H, Hr, k, IF archetype building. As a matter of fact, an initial progressive development of plastic hinges in beams is observed until the occurring of a column-sway mechanism (at the third story), leading to a final soft-story failure mode. For M-archetype buildings, a mixed-sway mechanism is also observed. In the final steps of the analysis, the plastic deformations are mainly concentrated in a single story, different for each case study building due to columns tapering (second story for M, Lr, k, IF, third story for M, Mr, k, IF, and fifth story for M, Hr, k, IF, respectively). A similar behavior is observed for L-archetype buildings. However, larger over strength ratio (OSR around 1.5) and ductility capacity (on average $\mu_c = 2.3$) have been obtained with respect to those observed for the corresponding M-archetypes (OSR $\approx$ 1.3 and $\mu_c \approx$ 1.8).

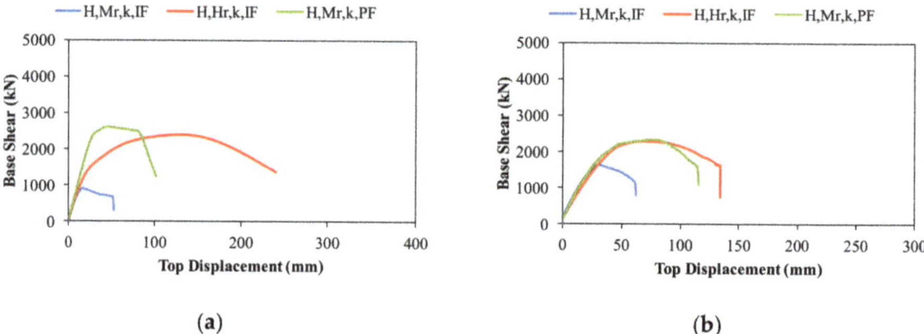

Figure 5. Pushover curves of the selected H-archetype buildings in (**a**) X-direction and (**b**) Y-direction.

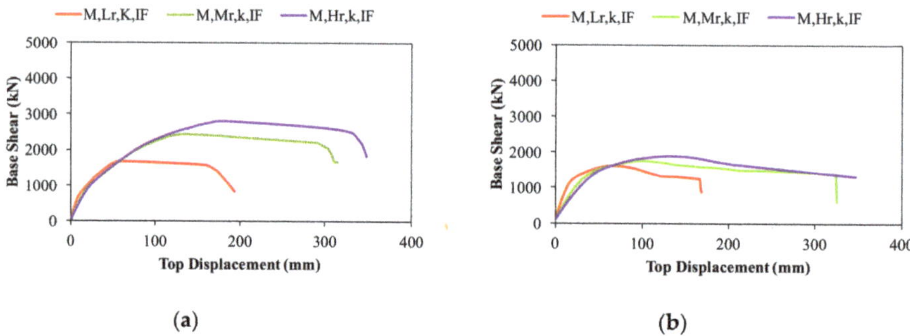

Figure 6. Pushover curves of the selected M-archetype buildings in (**a**) X-direction and (**b**) Y-direction.

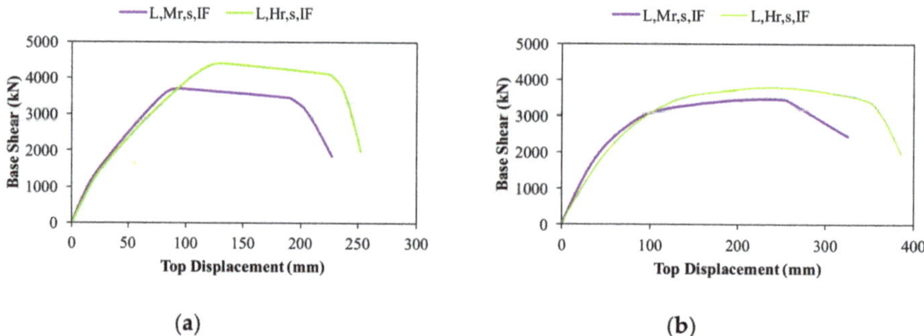

Figure 7. Pushover curves of the selected L-archetype buildings in (**a**) X-direction and (**b**) Y-direction.

Figures 8–10 show the maximum inter story drift profiles (calculated as average between the two principal directions) at the reference limit states (i.e., ZL, OP, and DC) for the H-, M-, and L-archetype buildings, respectively.

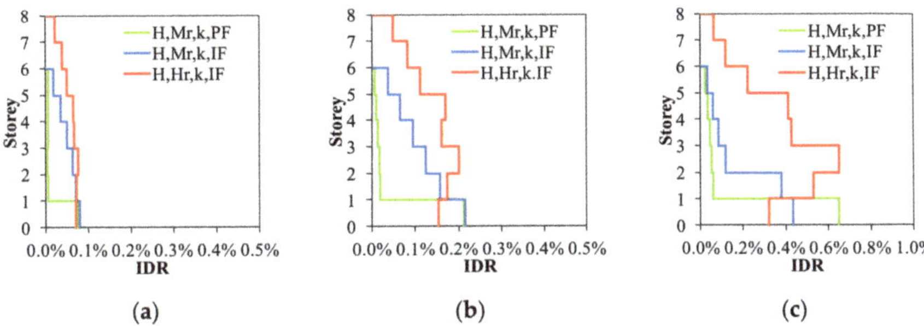

Figure 8. Maximum interstory drift profiles (computed as average between the two principal directions) at (**a**) ZL, (**b**) O, and (**c**) DC performance levels for the H-archetype buildings.

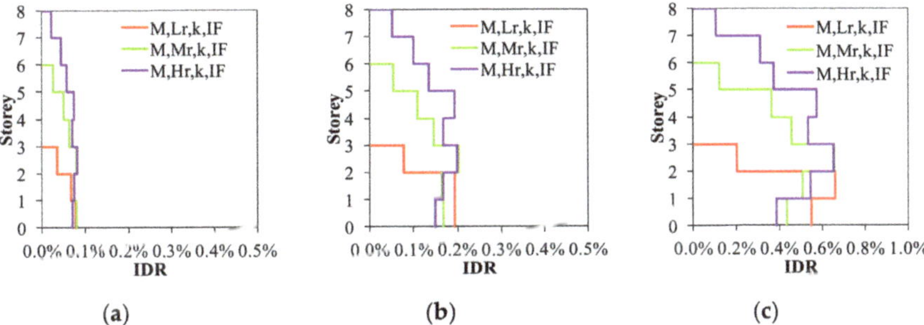

Figure 9. Maximum interstory drift profiles (computed as average between the two principal directions) at (**a**) ZL, (**b**) O, and (**c**) DC performance levels for the M-archetype buildings.

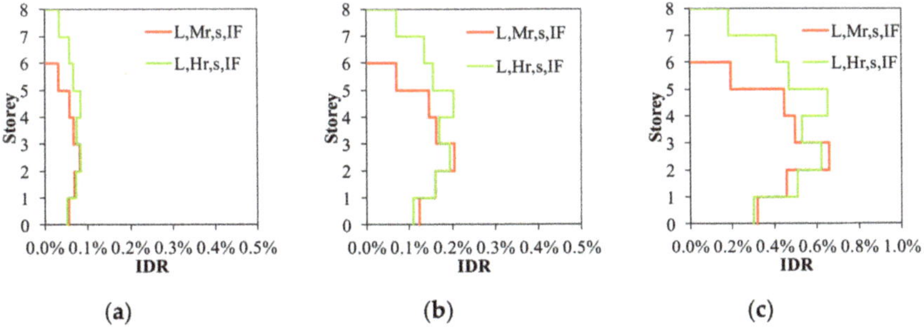

Figure 10. Maximum interstory drift profiles (computed as average between the two principal directions) at (**a**) ZL, (**b**) O, and (**c**) DC performance levels for the L-archetype buildings.

Generally speaking, the inter-story drift profiles show a bulged shape with larger values at the lower and mid stories, increasing with the seismic intensity (from ZL to DC), except for the case studies H, Mr, k, IF and H, Mr, k, PF. As a matter of fact, in these cases, the development of drift (hence damage) is concentrated in the lower stories. This is trivial for the PF case study, considering the effective structural configuration at first story. For what concerns the H, Mr, k, IF archetype, the described condition is related to the limited dimensions of the columns' section (30 × 30 cm continuously along the height of the building) together with the reduced strength and stiffness of the infill panels at the first story (due to large openings).

5. Expected Annual Loss Estimation

The DEAL approach has been performed as described in Section 2, considering three reference limit states, namely ZL, OP, and DC, and adopting suitable story-based loss functions (see Figure 11) derived by Cardone et al. (2020) [4] for first story, typical stories, and top story, assuming a unit RepC equal to €865/m^2 (see Table 9).

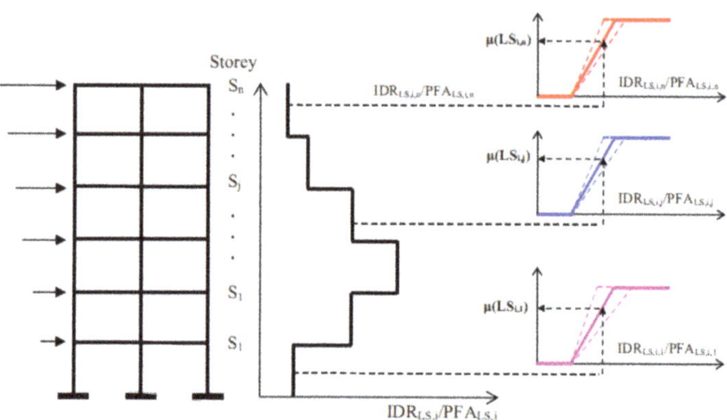

Figure 11. Story-based loss functions: example [3].

Table 9. Story-based loss functions: definition parameters.

Story	Destination of Use	Unit RepC €865/m^2	
		IDR$_{in}$	IDR$_{fin}$
First Story	Pilotis-type	0.05%	1.50%
First Story	Partially Infilled	0.05%	1.00%
First Story	Fully Infilled	0.05%	0.8%
Typical Story	Residential	0.05%	0.60%
Top Story	Residential	0.05%	0.40%

Table 10 summarizes the main outputs of the DEAL approach for the examined archetype buildings. In particular, the expected losses (μ) at the selected limit states and the corresponding spectral acceleration ($S_a(T^*)$) are shown. In Table 11, the effective values of direct EAL (EAL$_R$) in the as-built configuration are provided. Moreover, the values of the indirect component of EAL (EAL$_D$) are resumed, considering both the LB and UB conditions. The total values of EAL (EAL$_R$ + EAL$_D$), in the LB and UB, are also provided for each building typology.

Table 10. Main results obtained using the DEAL approach.

Building Typology	T* (s)	$S_{a,ZL}$ (T*) (g)	$S_{a,OP}$ (T*) (g)	$S_{a,DC}$ (T*) (g)	μ_{ZL} (%RepC)	μ_{OP} (%RepC)	μ_C (%RepC)
L, Hr, s, IF	0.95	0.046	0.104	0.291	3.11	17.91	72.76
L, Mr, s, IF	0.73	0.056	0.129	0.375	2.30	16.84	67.01
M, Hr, k, IF	1.02	0.041	0.090	0.227	2.75	16.75	67.69
M, Mr, k, IF	0.77	0.05	0.107	0.286	2.65	15.52	64.36
M, Lr, k, IF	0.42	0.066	0.178	0.407	2.20	17.73	70.17
H, Hr, k, IF	0.78	0.063	0.135	0.317	2.15	15.30	50.60
H, Mr, k, PF	0.99	0.015	0.036	0.114	0.30	1.85	7.19
H, Mr, k, IF	0.61	0.068	0.119	0.235	1.63	10.86	21.94

Table 11. EAL values derived from DEAL approach.

Building Typology	EAL_R (%RepC)	$EAL_{R,retrofit}$ (%RepC)	$EAL_{D,LB}$ (%RepC)	$EAL_{D,UB}$ (%RepC)	$EAL_{T,LB}$ (%RepC)	$EAL_{T,LB}$ (%RepC)
L, Hr, s, IF	1.76	1.00	0.14	0.22	1.9	1.98
L, Mr, s, IF	1.68	1.00	0.13	0.21	1.81	1.89
M, Hr, k, IF	1.96	1.00	0.17	0.26	2.13	2.22
M, Mr, k, IF	2.03	1.00	0.16	0.25	2.19	2.28
M, Lr, k, IF	2.75	1.50	0.19	0.31	2.94	3.06
H, Hr, k, IF	1.3	0.50	0.17	0.26	1.47	1.56
H, Mr, k, PF	0.62	0.50	0.08	0.12	0.7	0.74
H, Mr, k, IF	1.27	0.50	0.16	0.25	1.43	1.52

The results obtained in terms of EAL_R are directly related to the structural response of the examined buildings during the non-linear analyses. Indeed, the effective shape of the drift profiles directly affects the distribution of damage (and monetary losses) along the height of the buildings. For the H-archetypes, EAL_R values of the order of 1.3% have been obtained in the Infilled Frame configuration (H, Hr, k, IF and H, Mr, k, IF), while a lower value has been obtained for the Pilotis Frame archetype building (H, Mr, k, PF). As a matter of fact, the concentration of drifts (hence damage and losses) at the first story of the PF archetype significantly reduces the value of EAL_R. Similarly, the almost uniform distribution of IDR and damage observed for the M-archetypes produces larger values of EAL_R, of about 2.25% on average. This consideration is somehow emphasized for the M, Lr, k, IF archetype, where the IDR values appears constant for the 2/3 of the building height, leading to the peak value of EAL_R (2.75%). Finally, normalized spectral acceleration $(S_a(T^*)/S_a(T^*)_{ZL})$ being equal, the enhanced performances of the L-archetypes determinates a reduction of the direct monetary losses with respect to M-archetypes, leading to lower values of EAL_R (of approximately 25%, on average). Values of EAL_D of the order of 0.15% and 0.25% have been obtained for the LB and UB condition, respectively. Considerations similar to those reported for the direct component of EAL can be made with reference to the EAL_D values obtained for each macro-typology (H-, M-, and L-). Generally speaking, the contribution of the indirect component of EAL is of the order of 15% (10%) in the UB (LB) condition.

In order to have a general understanding of the potential socio-economic effects of the seismic scenarios at territorial scale, the EAL_T values obtained for the specific archetype buildings have been monetized with respect to the actual replacement cost. Subsequently, the monetary expected annual loss of the two compartments has been calculated by multiplying, for each building typology, the EAL_T value for the total number of elements of this category. As reported in Table 12, the total expected annual loss of the entire "virtual" compartments (i.e., area's expected annual loss, AEAL) amounts to approximately 540 millions of Euro (average between LB and UB condition). To give some reference for the mentioned value, the latter can be compared to the area's total income (ATI) obtained multiplying the per capita income (PCI) measured in the examined area in the last year (equal to €16,522 [22]), by the total number of occupants (equal to 4330). All that considered, the ratio between AEAL and ATI is equal to approximately 7.5 (540 millions divided by 71.5 million Euros). In other words, the average economic loss that is expected to accrue every year in the examined area, considering both the repair costs and the indirect costs related to downtime, is seven times larger than the annual average income of all the inhabitants. In practice, if only private resources were used to recover the economic losses and assuming that each owner gets a mortgage with an annual payment equal to 1/5 of his PCI (approximately €3300), 40 years are needed to pay back the economic losses accrued in only one year.

3. Vona, M.; Manganelli, B.; Tataranna, S.; Anelli, A. An optimized procedure to estimate the economic seismic losses of existing reinforced concrete buildings due to seismic damage. *Buildings* **2018**, *8*, 144. [CrossRef]

4. Cardone, D.; Perrone, G.; Flora, A. Displacement-Based Simplified Seismic Loss Assessment of Pre-70S RC Buildings. *J. Earth. Eng.* **2020**, *24*, 82–113. [CrossRef]

5. Porter, K.A. An overview of PEER's performance-based earthquake engineering methodology. In Proceedings of the Ninth International Conference on Applications of Probability and Statistics in Engineering, San Francisco, CA, USA, 6–9 July 2003.

6. Porter, K.A.; Beck, J.L.; Shaikhutdinov, R.V. Simplified estimation of economic seismic risk for buildings. *Earth. Spectra* **2004**, *20*, 1239–1263. [CrossRef]

7. Krawinkler, H. *Van Nuys Hotel Building Testbed Report: Exercising Seismic Performance Assessment*; Pacific Earthquake Engineering Research Center College of Engineering University of California: Berkeley, CA, USA, 2005.

8. Anelli, A.; Hidalgo, S.S.C.; Vona, M.; Tarque, N.; Laterza, M. A proactive and resilient seismic risk mitigation strategy for existing school buildings. *Struct. Infrastruct. Eng.* **2019**, *15*, 137–151. [CrossRef]

9. Vona, M. Proactive Actions Based on a Resilient Approach to Urban Seismic Risk Mitigation. *Open Constr. Build. Technol. J.* **2020**, *14*, 321–335. [CrossRef]

10. Sullivan, T.J.; Calvi, G.M. Considerations for the seismic assessment of buildings using the direct displacement-based assessment approach. In Proceedings of the ANIDIS Conference, Bari, Italy, 18–22 September 2011.

11. Ramirez, C.M.; Miranda, E. *Building Specific Loss Estimation Methods Tools for Simplified Performance Based Earthquake Engineering*; Technical Report No. 171; John A. Blume Earthquake. Engineering Center: Palo Alto, CA, USA, May 2009.

12. *Linee Guida per la Classificazione del Rischio Sismico delle Costruzioni*; numero 58 del; Decreto Ministeriale: Rome, Italy, 2017.

13. Perrone, G.; Cardone, D.; O'Reilly, G.J.; Sullivan, T.J. Developing a direct approach for estimating expected annual losses of Italian buildings. *J. Earth. Eng.* **2019**, 1–32. [CrossRef]

14. Cardone, D.; Sullivan, T.J.; Gesualdi, G.; Perrone, G. Simplified estimation of the expected annual loss of reinforced concrete buildings. *Earth. Eng. Struct. Dyn.* **2017**, *46*, 2009–2032. [CrossRef]

15. McKenna, F. OpenSees: A Framework for Earthquake Engineering Simulation. *Comput. Sci. Eng.* **2011**, *13*, 58–66. [CrossRef]

16. Cardone, D.; Flora, A.; Picione, M.D.L.; Martoccia, A. Estimating direct and indirect losses due to earthquake damage in residential RC buildings. *Soil Dyn. Earthq. Eng.* **2019**, *126*, 105801. [CrossRef]

17. Vamvatsikos, D. Derivation of new SAC/FEMA performance evaluation solutions with second-order hazard approximation. *Earthq. Eng. Struct. Dyn.* **2013**, *42*, 1171–1188. [CrossRef]

18. *Norme Tecniche per le Costruzioni*; NTC2018; Ufficio Pubblicazione Leggi E Decreti: Rome, Italy, 2018.

19. Cardone, D.; Perrone, G. Damage and Loss Assessment of Pre-70 RC Frame Buildings with FEMA P-58. *J. Earthq. Eng.* **2016**, *21*, 1–39. [CrossRef]

20. Applied Technology Council. *Next-Generation Seismic Performance Assessment for Buildings*; FEMA P-58-1 Federal; Emergency Management Agency: Washington, DC, USA, 2012.

21. Cardone, D.; Flora, A. Multiple inelastic mechanisms analysis (MIMA): A simplified method for the estimation of the seismic response of RC frame buildings. *Eng. Struct.* **2017**, *145*, 368–380. [CrossRef]

22. Italian National Statistics Institute (ISTAT). *15th National Census on Buildings and Population*; Italian National Statistics Institute (ISTAT): Rome, Italy, 2014. (In Italian)

23. Gizzi, F.T.; Masini, N. Historical earthquakes and damage patterns in Potenza (Basilicata, Southern Italy). *Ann. Geophys.* **2007**, *50*, 676–687.

24. Chiauzzi, L.; Masi, A.; Mucciarelli, M.; Vona, M.; Pacor, F.; Cultrera, G.; Gallovič, F.; Emolo, A. Building damage scenarios based on exploitation of Housner intensity derived from finite faults ground motion simulations. *Bull. Earthq. Eng.* **2011**, *10*, 517–545. [CrossRef]

25. Grunthal, G. *European Macroseismic Scale (EMS–98)*; Cahiers du Centre Européen de Géodynamique et de Séismologie: Luxembourg, 1998.

26. Dolce, M.; Kappos, A.; Masi, A.; Penelis, G.; Vona, M. Vulnerability assessment and earthquake damage scenarios of the building stock of Potenza (Southern Italy) using Italian and Greek methodologies. *Eng. Struct.* **2006**, *28*, 357–371. [CrossRef]

27. Brzev, S.; Scawthorn, C.; Charleson, A.W.; Jaiswal, K. *Interim Overview of GEM Build-ing Taxonomy V2.0*; GEM Technical Report Version 1.0; GEM Foundation: Pavia, Italy, 2013.

28. Flora, A.; Iacovino, C.; Cardone, D.; Vona, M. *Typological Inventory of Residential Reinforced Concrete Buildings for the City of Potenza*; Lecture Notes in Computer Science: Cagliari, Italy, 2020; pp. 899–913.

29. Masi, A.; Vona, M. Vulnerability assessment of gravity-load designed RC buildings: Evaluation of seismic capacity through non-linear dynamic analyses. *Eng. Struct.* **2012**, *45*, 257–269. [CrossRef]

30. Collegio degli ingegneri e degli architetti di Milano. *Prezzo Tipologie Edilizie. 2014*; Tipografica del Genio Civile: Rome, Italy, 2014. (in Italian)

31. Vona, M. A Review of Experimental Results about In Situ Concrete Strength. *Adv. Mater. Res.* **2013**, *773*, 278–282. [CrossRef]

32. Masi, A.; Chiauzzi, L. An experimental study on the within-member variability of in situ concrete strength in RC building structures. *Constr. Build. Mater.* **2013**, *47*, 951–961. [CrossRef]

33. Applied Technology Council. *Commentary on the Guidelines for the Seismic Rehabilitation of Buildings*; FEMA 274-NEHRP; Federal Emergency Management Agency: Washington, DC, USA, 1997.